Where Science and Ethics Meet

Where Science and Ethics Meet

Dilemmas at the Frontiers of Medicine and Biology

~

Chris Willmott and Salvador Macip

PRAEGER™

An Imprint of ABC-CLIO, LLC

Santa Barbara, California • Denver, Colorado

Library of Congress Cataloging-in-Publication Data

Names: Willmott, Chris, author. | Macip, Salvador, author.
Title: Where science and ethics meet : dilemmas at the frontiers of medicine
    and biology / Chris Willmott and Salvador Macip.
Other titles: Jugar a ser déus. English
Description: Santa Barbara, California : Praeger, [2016] | Includes bibliographical
    references and index.
Identifiers: LCCN 2016003437| ISBN 9781440851346 (alk. paper) |
    ISBN 9781440851353 (eisbn)
Subjects: LCSH: Science—Moral and ethical aspects.
Classification: LCC Q175.35 .W55 2016 | DDC 174/.95—dc23 LC record
    available at https://lccn.loc.gov/2016003437

ISBN: 978-1-4408-5134-6
EISBN: 978-1-4408-5135-3

20   19   18   17   2   3   4   5

This book is also available on the World Wide Web as an eBook.
Visit www.abc-clio.com for details.

Praeger
An Imprint of ABC-CLIO, LLC

ABC-CLIO, LLC
130 Cremona Drive, P.O. Box 1911
Santa Barbara, California 93116-1911

This book is printed on acid-free paper ∞
Manufactured in the United States of America

I consider ethics to be an exclusively human concern with no superhuman authority behind it.

Albert Einstein

Whenever science makes a discovery, the devil grabs it while the angels are debating the best way to use it.

Alan Valentine

Things change so fast, you can't use 1971 ethics on someone born in 1971.

Grace Slick

Contents

Preface

Even a casual follower of news broadcasts will be aware of the frenetic pace with which new advances in biology and biomedicine are being reported. It seems that barely a week goes by without the announcement of groundbreaking discoveries. Alongside reflection on the science, it is also important to consider the social consequences that would follow any of these innovations.

With each advance that has the potential to improve our lives comes a series of important changes that could alter forever the norms of our civilization. Extending the lifespan beyond 100 years. Human cloning. Regenerative therapies. Genetic manipulation. Prenatal screening. New cures for old diseases and an end to suffering for lots of people. The amazing gifts that science is promising may come with a high price tag. All consequences need to be carefully considered if we want to avoid entanglement in a dystopian future.

In this book, we will focus on some of the most exciting areas of development in biology and medicine. Each chapter will start with a short scenario to set the scene regarding one or more applications of the innovative technology. At the end of the case study we will invite the reader to pause and consider the moral issues that derive from it. After this, we will try to present a balanced account of both the current state of the science and the ethical implications of each new development. We will close each chapter with a summary of arguments in favor of and against the changes that each biomedical advance is likely to bring.

Our goal is to encourage readers to think and eventually choose their own point of view. This is why we have tried to include sufficient pieces of the puzzle to allow an informed decision to be made. Too often we are just aware of one side of any complex story, and this has a negative impact on the decisions we make. We think it is always important to expand our knowledge before committing to an opinion.

It will soon become necessary for all of us to reflect on the matters discussed in this book. Only those who understand the moral and social implications of the new scientific discoveries will be able to participate in the debate and help decide which way humanity will go in the future. It's never too early to start.

CHAPTER 1

Designer Babies:
Choosing Our Children

A BUSY MORNING AT THE CLINIC

"Good Morning, Jackie, can you please bring me the appointments diary through to the consultation room?"

Professor Simon Fraser knew it was rather old-fashioned to use a paper diary rather than some electronic version. He'd been struck more than once by the irony of this low-tech preference, given the cutting edge treatments he provided for his clients. Nevertheless, he liked to be able to flick through the pages and see how the week was shaping up.

Fraser had established the Randolph Fertility Clinic in the early 1990s, and by any measure it had been a hugely successful venture. The clinic had one of the highest conception rates for standard **in vitro fertilization** (IVF) and a reputation for introducing groundbreaking new procedures. Their PR agent made sure the media was well briefed about these pioneering treatments, and regular news coverage had turned Fraser into a minor celebrity. His rugged good looks and easy-going manner had added to his TV friendly credentials, and he got regular invitations to appear on chat shows. The Porsche and the holiday home in Tuscany were also fruits of his labors.

Jackie came through to the clinic from reception bringing the diary and Fraser's cup of coffee—he didn't need to ask for the latter; their morning routine had been the same for the last 15 years.

"Thanks, Jaq," said Fraser.

He took the diary and the coffee from his secretary and settled into his leather desk chair. He saw that he had four consultations lined up this morning. Fraser already knew two of the couples.

The Jamesons had previously been through two cycles of IVF with the clinic, and although they had apparently produced high-quality embryos, Mrs. Jameson had miscarried during the first three months on both occasions. Today they were coming in to discuss having another go. Fraser was going to suggest that they try **Preimplantation Genetic Screening** this time. It was a relatively new service offered by the clinic, in which they checked that the embryos had the correct number of chromosomes before putting one back into the mother's uterus. It might turn out to be the crucial factor in helping Mrs. Jameson carry her baby to term: there had been some recent reports showing that a significant percentage of miscarriages were in fact due to chromosomal abnormalities.

Nikos and Maria Dimitriou had also been to the clinic before, and had brought their daughter Eleni with them. From a young age Eleni had been pale, prone to infections, and, in the words of her pediatricians, generally "failed to thrive." Tests showed that she was suffering from beta-thalassemia—a blood disorder caused by an inherited mutation in one of her genes. Without realizing it, both of Eleni's parents were **carriers** for the condition, each having one faulty copy and one good copy of the gene. Since they each possessed this good copy, Maria and Nikos did not show any symptoms themselves, but the laws of genetics stated that there was now a one in four chance that a second child would have the same bad luck as Eleni, inheriting faulty copies of the gene from both parents. The Dimitrious didn't want to take this risk, so they were asking Fraser to make sure that the embryo would not carry the genes for thalassemia, before implanting it into Maria's uterus.

This analysis, called **Preimplantation Genetic Diagnosis**, or PGD for short, was a fairly typical request. In the case of the Dimitrous, however, they wanted to add a second test. They wanted to screen their embryos to make sure that the new child would be a "tissue match" for Eleni. That is to say, genetically close enough to be a potential donor. Assuming this was possible, they hoped to harvest stem cells from the umbilical cord when the new baby was born. They could then have the cells transplanted into Eleni, in order, hopefully, to cure her disease. This was known as a **savior sibling** procedure.

VOCABULARY

Assisted reproduction: a series of techniques aimed at helping couples to get pregnant, including artificial insemination and IVF.

Carrier: a person is a carrier if he or she has only one defective copy of the gene for which two faulty copies are needed to cause a particular disease. As such the person will not suffer from the illness themselves, but run the risk of passing it on to their children.

In vitro fertilization (IVF): currently the most common system of assisted reproduction. With this technique, eggs are fertilized by sperm in a petri dish and are incubated there for a few days while their development is monitored. Then they are implanted into a uterus.

Preimplantation genetic diagnosis (PGD): genetic analysis of an embryo before it is going to be implanted in a uterus in a procedure of assisted reproduction. It can be used to check for common specific genetic disorders that cause diseases, but it can be used also to choose the sex of the embryo or a "savior sibling."

Preimplantation genetic screening (PGS): a form of PGD in which the embryos are checked for major problems in the structures of their chromosomes.

Savior sibling: term used to define a child conceived via IVF as a potential tissue donor for a sick brother or sister. PGD is performed to determine whether the embryo is a genetic match for the recipient of the transplant.

He hadn't met the third family, but Fraser was aware of their circumstances as he'd read about it in the local press. The previous November there had been a terrible fire at the Moores' house. Mum and Dad had escaped along with their three daughters, but their twin sons, sleeping in a top floor bedroom, had perished. It was widely reported at the time that the Moores were determined to have at least one more son to "restore the balance in their family" as Pat Moore had put it in one interview. Fraser felt sure their visit today was the first step in fulfilling that intention.

The reason for the consultation with the fourth couple, Dean Blackstock and Tracey Rollins, was not clear, but Fraser would know soon enough, since they were actually his first appointment for the day and would be here in a few minutes.

Fraser put the diary to one side and sipped his coffee. A couple of moments later, there was a buzz on the intercom. Jackie announced that Mr. Blackstock, Ms. Rollins, and another woman were here. Odd, thought Fraser. There had been no mention of a third visitor. He got up from his chair and moved over toward the door to greet his clients, just as they were being shown in by Jackie.

"Good morning, Professor Fraser," said the first woman while offering her right hand. "I'm Mary Scott and I'll be translating for Dean and Tracey today."

Noting the bemused look on Fraser's face, she continued.

"Dean and Tracey are profoundly deaf—they can lip-read anything you say, but I'm here to ensure that you are able to understand them."

The other two visitors now stepped forward and shook Fraser's hand.

"Hello," said Dean with the muffled intonation typical of deaf people.

Fraser returned the greeting and indicated towards the sofas where he tended to conduct his consultations. When they had all sat down, he spoke once more. His head was turned towards Dean and Tracey as he asked how he could be of assistance, but his eyes were directed towards Mary Scott.

Dean responded first, speaking out loud, but Fraser essentially understood nothing of what was being said. However, Dean was simultaneously using sign language and, after a few moments, paused quite naturally for Mary to clarify everything.

"Dean and Tracey have both been deaf since they were born. They've been living together for the past three years and now they want to start a family. They've read that it is possible to use genetic testing to know whether or not a baby is going to be deaf."

So, the mystery visitation was another request for PGD. Like the Dimitriou family and their desire to ensure that the new child would not have thalassemia, it seemed Dean and Tracey wanted to make sure that they would not pass whatever "deafness genes" they had to their offspring. Fraser felt on familiar ground now and started to speak confidently to the couple and their colleague.

"You're absolutely right," he began. "It is possible for us to carry out some genetics tests as part of in vitro fertility treatment. There are several standard tests for common diseases, but the range of genes that we can screen for is increasing all the time. Indeed, a number of genes have been identified that can cause deafness. We can be fairly confident of producing a child that does not share their disability."

At this point it was Tracey who started to sign with some intensity. Mary nodded and turned to Fraser.

"I'm afraid, Professor Fraser, that you've misunderstood their request; Tracey and Dean want to make sure that their child *is*d eaf."

THINK ABOUT IT . . .

If Dean and Tracey had been asking to *avoid* their baby being deaf, would this have been an appropriate use of genetic screening?

Does the fact that they want to make sure the child *is* deaf make any difference to whether this is right or wrong?

Think about the other clients due to see Professor Fraser that morning. They were coming to the clinic for a variety of different reasons—to try to have a healthy baby, to produce a "savior sibling" to help an older child, and to choose the sex of their next baby. Do you agree or disagree with these uses of IVF?

WITH A LITTLE HELP FROM MY DOCTORS

All of the patients talking to Professor Fraser in our scenario were seeking various applications of a technique known as Preimplantation Genetic Diagnosis.[1] PGD is, however, entirely reliant on the underlying developments of *in vitro* fertilization (IVF). It is therefore important that we begin by reflecting on some of the scientific and ethical issues associated with IVF itself before moving on to consider PGD in more detail.

Most people assume that when the time comes to start a family, nature will simply kick in. Take one egg, add sufficient sperm to ensure one achieves fertilization, and provide a cozy uterus for nine months of incubation. For a growing number of couples, however, the reality is far more traumatic. For whatever reason, they have difficulty achieving or maintaining a pregnancy.

Until the 1970s, this was a situation that they could do little to alter. Then in July 1978, Louise Brown was born. Louise was the first "test tube baby," the fruit of pioneering work on IVF by Patrick Steptoe and Robert Edwards. Robert Edwards received a Nobel Prize in 2010 for this work on **assisted reproduction** (Steptoe had died in 1988, and the Nobel Prize is not awarded posthumously).

THE YOUNGER BROTHER

In 1990, a couple underwent IVF treatment. They had five leftover embryos, which were frozen. Twenty years later, these embryos were donated to another couple that also had fertility problems. Only two of them survived after being thawed, but in October 2010, the 42-year-old woman receiving the donated embryos gave birth to a healthy child. This new baby therefore had a "twin" conceived at the same time but now 20 years older than him!

It is not known for how long frozen embryos remain viable. Twenty years is the maximum that has been tested so far, but it could be just the beginning. This poses an interesting ethical dilemma: a woman could in theory use an embryo frozen by her parents several decades before and give birth to a child that would genetically be her brother or sister.

In vitro (literally "in glass") fertilization of the egg is done outside the body, usually in a sterile plastic Petri dish. The fertilized egg is then incubated while it undergoes three or four rounds of cell division before it is implanted into a uterus, usually around three days after fertilization. These embryos can also be frozen and implanted at later stages (see box: The Younger Brother).[2] Alternatively they can be donated to other couples, although in reality "spare" embryos are often just discarded. This is one of the reasons some, for example the Catholic Church and evangelical Christian groups, object to IVF on ethical or religious grounds.

The controversy surrounding assisted reproduction goes even further. Thanks to this technique, it is easier for unmarried mothers or same-sex couples to have children. For instance, a lesbian couple can use a male donor to fertilize an egg through IVF. It has also been hotly debated whether postmenopausal women should be allowed to undergo IVF.[3] Some clinics have a limit of 50–55 years, but others do not apply age criteria so strictly, with the result that relatively old women can get pregnant, provided that they can obtain fresh eggs from a younger donor (see box: The Oldest Mothers in the World).[4]

There are a range of risks inherent in women having children so late in their own lives. Not only are there the risks associated with the pregnancy itself; there are the pressures of coping with toddlers and the danger that parents will be dead before the child reaches their own adulthood. Historically, of course, the death of a mother while her children are young has not

THE OLDEST MOTHERS IN THE WORLD

In December 2006 Maria del Carmen Bousada de Lara gave birth to twins at Sant Pau Hospital in Barcelona. It was a week before her 67th birthday. As the only daughter of her parents (she had three brothers), Maria had taken responsibility for the care of her mother at the expense of her own opportunities to marry and have a family. It was her great misfortune in this regard that her mother lived to the age of 101. After the death of her mother, Maria flew to Los Angeles where she received donated embryos having, it is alleged, lied about her age. Tragically, Maria died of cancer in 2009 leaving her twin boys as orphans.

Given the cultural stigma of childlessness in India, many of the cases involving postmenopausal women are taking place in that country. Both Rajo Devi and Omkari Panwar were 70 in 2008 when they gave birth to IVF children, with the latter having twins. The oldest mother of triplets is also from India. Bhateri Devi was 66 when she had three children in 2010.

been unusual. However, these newer technologies are making it *likely* rather than *possible*. As a result, many people strongly oppose IVF in cases of this kind. Others, however, defend the right of a woman to become a mother whenever she chooses, without being subjected to the usual biological limitations.

When IVF was being developed as a technique, the success rate was relatively low. Fertility clinics therefore got into the habit of implanting several embryos into a woman's uterus to try to ensure that one would take successfully. Of course there were times that more than one baby would develop, leading to overcrowding in the uterus and babies being born small and prematurely and at the risk of a lifetime of poor health. To guard against this, several countries have set a limit of two or even one embryo to be transferred to a woman during any one cycle of IVF.

Despite these safeguards, multiple births do still occur in IVF. A notorious case involved Nadya Suleman of Los Angeles.[5] In 2009 Ms. Suleman gave birth to octuplets. Very unusually, all eight implanted embryos developed and survived. Initially there was public support for this amazing feat, with the press giving her the nickname "Octomom." As more details of the case came to light, however, serious ethical questions have been asked. Not

only did it transpire that Suleman already had six other children, also conceived using IVF, it came to light that her doctor, Michael Kamrava, had in
fact transferred 12 embryos at one time. In June 2011, the Californian
medical board examining the case described this as an "extreme departure
from the standard of care" and revoked his license to perform fertility
treatments.[6]

Since IVF is a relatively new technique, there are still some concerns
about the health of children conceived this way. Could there be unexpected problems associated? When the activation and regulations of genes
in IVF babies has been compared with similar processes in naturally conceived babies, there have been some observed differences relating to perhaps 5 to 10 percent of the total number of genes in their genomes.[7] Does
this mean anything? Is assisted reproduction unexpectedly introducing
changes in embryos, the consequences of which we will not know for
many years?

Although it is now more than 30 years since IVF was invented, we cannot yet know for sure whether or not there will be long-term health issues
associated with the technique. Some of the first children born using IVF
methods have successfully had children of their own, but it remains a possibility that they may have greater predisposition to suffer from certain
diseases when they age.

HOW DO YOU CHECK THE DNA OF AN EMBRYO?

After sperm has been used to fertilize an egg, the resultant cell will
divide once, turning it into two cells, and then each of these will divide a second time to make four cells. Each of these divides a further
time to give eight cells, all sitting within a protective jelly coat. At this
stage of development, it seems that taking one of the cells away will
not cause long-term harm to the embryo. A hole is made in the jelly
and one cell is carefully removed for testing. In the meantime, the
other cells go on dividing and seem to cope without the missing cell.

PGD works on the basis that the things you find out about the tested
cell will be equally true for the other cells sitting inside their protective
jelly layer. Provided that everything about the tested cell looks correct,
the remaining collection of cells can then be transferred to the mother's uterus, where it is hoped that they will implant and continue to
develop.

THE AGE OF DESIGNER BABIES

The term *designer baby* is widely used in newspapers and television reports whenever the selection of features of a child is being discussed. Despite the fact that this expression has entered the popular culture, it is actually used in a slippery way that covers a broad range of technologies, some of which are real, some potential, some improbable and even some flat-out impossible. To what extent are we able to choose the physical and mental characteristics of our sons and daughters? Can we really manipulate the genes of a human embryo to improve it over and above normal selection?

Intervening in human reproduction has been the subject of speculation in science fiction on numerous occasions. Probably the most famous dystopia based on the complete control of all the steps in human reproduction is Aldous Huxley's *Brave New World*. In the novel, human population numbers are kept stable and the physical and mental characteristics of humans are determined by chemical treatment of embryos incubated in factories. More recently, the film *GATTACA* had depicted a near future world in which genetic selection of individuals has become routine, and genetically enhanced "Valids" lord it over the "Invalids" whose genomes have been assembled in the more time-honored manner.

Genetic manipulation of humans (the procedures required to deliberately introduce multiple characteristics into one individual), remains far beyond the capabilities of current science, as we will discuss in the next chapter, and is also specifically forbidden by law in most countries. Currently, we are limited to PGD as a way to select embryos with certain genetic characteristics in IVF (see box: First Steps).[8] It is important to remember that PGD itself can only select for or against genes that are already present in the genetic makeup of the parents, not add or remove anything. The Randolph Clinic is fictional, but the treatments for which the couples were seeking the help of Professor Fraser are all possible with existing technology, and will be increasingly familiar at fertility clinics worldwide over the next few years.

A BLESSING OR A CURSE?

PGD has become an established procedure in many countries around the world, but this should not disguise the fact that it remains highly controversial, irrespective of the particular motivations for carrying it out. Even within Europe, there are significant differences regarding permissible uses, with outright bans in some countries, including Austria, Ireland, and

> ## FIRST STEPS
>
> The first clinical use of PGD was actually reported in 1990, when a team at the Hammersmith Hospital in London examined IVF embryos to make sure that they did not have a Y chromosome, the tell-tale sign that the embryo was male. Because males only have one copy of the X chromosome, they are susceptible to "sex-linked" illnesses. These are not usually a problem for females because the second copy of the X chromosome that they possess is very unlikely to have the same fault and will be able to cover for any genetic defect.

Switzerland.[9] Germany was also among these countries, but in the summer of 2011 the federal parliament voted in favor of allowing it in some special cases, for example when there is the risk of passing on a known genetic disease.[10]

There are a variety of reasons why people object to the use of the technique at all. Some are concerned by the fate of embryos that are *not* selected, as we indicated before. After all, you go to all the trouble of screening embryos only to eventually reject the unsuitable ones. If you believe that life begins at conception, then you may well view the destruction of rejected embryos as morally wrong. Jay and Ami Bharvada, a couple from London, found a solution to this problem.[11] Like the Dimitrous in our story, they were looking for a savior sibling that could be a genetic match to their first son, Jivan, who has a rare disease called Wiscott-Aldrich syndrome. They finally succeeded, and Jaya was born in 2011. The Bharvadas are Jains, and their religion forbids them to cause harm to any living creature. They believe this includes embryos, so they decided that instead of getting rid of the 15 to 20 leftover embryos they produced, they would eventually implant them all. This means that, according to the typical success rate, the Bharvadas may end up with at least five further children, which may be a little taxing on them, but at least will not clash with their beliefs.

It is not only supporters of pro-life positions who wish to maintain an outright ban on PGD. The whole idea of "selecting" children raises objections from several quarters. First, there are those who object to selection on the grounds that it fundamentally alters the relationship between parent and child; a "gift" becomes a "commodity." The ability to choose characteristics of our children turns parenthood into a financial transaction with certain expectations regarding the "product" we have ordered.

A DIFFICULT CHOICE?

Ethicist John Harris presents an evolving series of scenarios to support his view that allowing PGD under some circumstances but not under others is nonsensical. Imagine, he asks, that a woman has six embryos available, with the intention to implant only three through IVF. If three embryos prove to be normal and three have genetic disease, there is a clear expectation that the three healthy embryos will be transferred. Now, however, assume that three are normal, but the other three will have longer, healthier lives—which three should you choose to implant? Finally, assume that three are normal, but three are diagnosed as having superior intelligence—which now should you implant? Harris reasons that if knowledge of this kind is available, we ought to make the most of it. Some see this as a modern rationalization of eugenics.

Second, a willingness to "discard" embryos facing genetic disabilities (such as deafness in our example) can be taken as a huge slur on individuals living with those conditions. Many disabled-rights activists argue that it is society's attitude to disability rather than the life of a "handicapped" individual that needs to be rejected. On the other hand, advocates for PGD counter that embryo selection makes no value statement regarding people already alive (see box: A Difficult Choice?).[12]

Third, objections to PGD are raised over the specter of eugenics, which is the attempt to influence the genetic make-up of the population by exerting control over reproduction. In less technological times, eugenics took the form of rules regarding who was allowed to procreate and sometimes included the forced sterilization of individuals deemed unsuitable as breeding stock. This movement is closely associated with the Nazi era in Germany, a precedent that played a role in the long-time ban on PGD in that country. However, at the end of the 19th century and the start of the 20th century eugenics had prominent supporters in many other countries, notably the United States, England, and Sweden.

Contemporary worries about the eugenic specter see the benign and well-meaning selection against genetic illnesses such as cystic fibrosis or beta-thalassemia as Trojan horses. Like the mythical wooden horse used by ancient Greeks to smuggle soldiers into the city of Troy, it is suggested that therapeutic uses of PGD open the way for other, more questionable reasons for selection: the deliberately picking of characteristics such as intelligence, sporting ability, or even sexuality, as soon as the genes involved in these features are finally known.

On the other hand, supporters of PGD would argue that freedom to make one's own reproductive choices is as much a mark of democratic society as freedom of thought or freedom of speech. For them, a choice to use PGD to make sure a child has a desired characteristic is, in essence, no different from a decision to use contraception or, indeed, about how many children to have.

ANY BABY WILL DO?

Different applications of PGD raise their own specific debates, which are illustrated by the circumstances bringing clients to see Professor Fraser at the Randolph Fertility Clinic during the morning described in our scenario. The simplest of all is the case of the Jamesons, the couple that had experienced several early miscarriages. They were not seeking to choose anything about their child. They would be overjoyed if the treatment were able to produce *any* baby for them. In situations like this, PGD screening would have no impact on the genetic characteristics of the child. It has lesser ethical implications than the other examples, beyond the fate of embryos that were not implanted.

Very different is the case of the Dimitrou family. One of the intended outcomes would be a new child free from the gene mutation that has caused thalassemia in their daughter, Eleni. First, therefore, they are intending to use PGD to ensure that the embryo does not carry the rogue gene and grows into a person without this particular inheritable condition.

Second, however, they intend to conduct an additional test to see if the new baby will be a suitable donor for Eleni, a step that raises fresh issues. There have already been several real-life cases where this dual-testing procedure has occurred. In the United States in 2000, baby Adam Nash was born as a savior sibling for his sister Molly.[13] Molly had suffered from another blood condition called Fanconi anemia. Stem cells from Adam's umbilical cord were successfully used to cure Molly, and at the 10th anniversary of the procedure, both were reported to be well.

In the United Kingdom, the production of savior siblings has been a thorny issue for the Human Fertilisation and Embryology Authority (HFEA), the body that regulates all fertility treatments. In a protracted test case, the HFEA initially gave permission for the parents of Zain Hashmi, who has thalassemia, to use PGD to produce a savior sibling. Campaigners led by a prominent pro-life group obtained a ban on the treatment, arguing

DO YOU LOVE ME FOR WHO I AM?

Savior siblings may face psychological difficulties. How will they feel later in life, knowing that they were produced for a particular purpose? This pressure would be compounded if, for whatever reason, the treatment failed. A lack of success with umbilical cord stem cells could lead to an escalation in the number of procedures the younger child would need to have in order to try to help their older sibling. Umbilical stem cell use might be followed by bone marrow transplantation, a painful and invasive procedure, and even the donation of an entire organ.

Readers familiar with Jodi Picoult's novel *My Sister's Keeper* or the 2009 film adaptation will recognize the playing out of this situation. In that story, Anna Fitzgerald was conceived as a savior sibling for her sister Kate, who suffers with leukemia. Treatments, however, have not gone well and at the point where Picoult introduces us to the Fitzgerald family, 13-year-old Anna is seeking legal emancipation from her parents to avoid being required to give a kidney to Kate.

that the HFEA had acted beyond its statutory limits in agreeing to this procedure. The heart of the issue was the question of who would benefit from the procedure. The HFEA argued that the savior sibling procedure sought by the Hashmis was appropriate since it benefitted both the older boy *and* the new child, who would be free from thalassemia. Ultimately, the High Court found in favor of the Hashmis and the HFEA[14] (although, as an interesting reminder that permission is not an automatic guarantee of success, the Hashmis have been unsuccessful in producing a suitable donor embryo).

At around the same time, the HFEA decided *not* to allow another couple, Michelle and Jayson Whitaker, to use PGD to help their son Charlie. Superficially, the cases looked identical. Charlie was also suffering from a blood disorder, this time a condition called Diamond-Blackfan anemia. Unlike thalassemia, however, the exact molecular cause of this anemia is not certain, and many cases are thought to be spontaneous rather than inherited. Because there is no clear test for it, it was argued that the Whitakers' case would benefit the older boy, but offered no net advantage to the new baby. In fact, given the additional procedures through which the embryo would have to be put to do the tissue testing, there was an overall increase in the risk burden for the younger child. The HFEA therefore said no.[15]

What happened next is illustrative of one of the difficulties arising from variation in legislation across different countries. Having been told that they could not have PGD on these grounds within the United Kingdom, the Whitakers became an early example of a phenomenon known as "fertility tourism"—they simply travelled to another country where they could have the treatment.[16] In their case, the Whitakers flew to Chicago, where a tissue-matched embryo was implanted. Nine months later Mrs. Whitaker gave birth to Jamie, and his stem cells were successfully used to treat Charlie.[17]

As it happens, the HFEA has subsequently revised its rules regarding use of tissue-typing to find a donor for a child with Diamond-Blackfan anemia.[18] There remain, however, a variety of procedures that are permitted in one country but not in another. Fertility tourism will therefore be likely to remain an important consideration for several years to come.

BEYOND ILLNESS: THE CUSTOM-MADE CHILD

One step beyond these situations is screening for physical characteristics that have nothing to do with any disease. We still don't know exactly how features like height, eye, skin and hair color, resistance to certain diseases, intelligence, or lifespan are determined. It is almost certain that many of these complex traits will be strongly influenced by environmental factors as well as our genes. Future advances in our understanding of genetics will no doubt offer valuable insights, but currently there are limited physiological factors that we can recognize (and thus select for) in a genetic screening.

The most obvious characteristic that we *can* select is a baby's gender. Following their heartbreaking accident, the Moore family in our scenario were determined that any further child they may have must be a boy. There are real-life echoes of their experience. In 1999, Nicole Masterton died in an accident.[19] Her parents, Alan and Louise, had four other children, but all were boys. They wanted to have a new daughter and sought permission from the HFEA. Another couple, Mike and Nicola Chenery, had not experienced tragedy but they too approached the HFEA for permission to have PGD to ensure that they could have a girl to join their existing family of four sons.[20]

In both cases, the HFEA said no. Sex-selection in the United Kingdom is only allowed in order to guard against X-linked genetic diseases. Wanting to choose the gender of a child as a result of disaster or for reasons of personal preference is not permitted. Barred from seeking PGD in the United Kingdom, both the Mastertons and the Chenerys took matters into their own hands. Like the Whitaker family discussed earlier, both couples traveled overseas to receive treatment. Mrs. Chenery went to Spain and now has two daughters. The Mastertons went to Italy but were only successful in producing male embryos, which they gave away.

Is there anything wrong in choosing the sex of your baby? Given that IVF allows us to do so without having to resort to an abortion, which would be the only way to deal with it in normal conception, why see it as a negative thing? We are not harming the child in any way, right? China's one-child policy has a bearing on this argument. We will discuss this further in a later chapter. At this point, however, it is sufficient to note that significant social problems arise from cultural preferences that often see having a male child as better than having a girl. Even before the rise of PGD, the population of children in regions of China and India was already heavily skewed towards boys, implying that selective abortion and even infanticide of female babies may be occurring. A report published by the World Health Organization in 2011 estimates that the ratio in some regions of central Asia is now 130 boys to every 100 girls.[21] If PGD for sex selection were allowed in these countries, the statistics might show even greater bias towards male children. The problem could extend also to other cultures depending on fashion and trends (see box: Was It Our Fault?).[22] The supernumerary male children will face future difficulties of their own, for example the shortage of potential wives. Considering that this puts the male child in severe disadvantage, there are strong arguments for preventing sex selection from becoming widely available.

WAS IT OUR FAULT?

Some studies blame the Western world for encouraging the current sex bias epidemic in countries like China and India. It all started in the 1960s, when a campaign for introducing safer abortion techniques and ultrasound machines in Asia was financed by the Rockefeller Foundation and the International Planned Parenthood Federation, among others. This was supposed to facilitate reproductive choices in the poorer regions and help control the population boom. A side effect was that it helped screen for female fetuses, which could now be easily aborted.

AND ONE MORE TURN OF THE SCREW . . .

Choosing a baby's sex is not the only decision that could potentially have a negative impact on a child. The case of Dean and Tracey in our scenario shows us that PGD could be used to specifically select for features that most people would consider detrimental. In what might be seen as the opposite of eugenics, they were actively seeking to have a child *with* a defect, one who is deaf and will fit comfortably into their parents' nonhearing world. Can we really consider that we are harming the child by making sure he will be born with a disability, or is this just a matter of parent's choice?[23]

We have to keep in mind that to do so, they would be actively rejecting any healthy embryos, which some would see as an absurd and clearly unethical proposition. Can it really be acceptable to get rid of an embryo just because it *doesn't* have a disability? Others might argue, however, that this is no different to the discarding of "leftover" embryos generated for IVF but no longer needed once the couple had successfully conceived all the children they wanted.

Discussing the deliberate selection of a deaf child over a hearing baby may sound like a philosopher's "thought experiment" to test the theoretical limits of PGD. However, in a 2006 survey of 190 American fertility clinics, 3 percent said that they had used PGD to intentionally select for a child with a disability, including deafness.[24] Similarly, British couple Tom and Paula Lichy asked for a deaf child when they started IVF treatments in 2008.[25]

Members of the deaf community frequently balk at calling deafness a disability, tending instead to see it as delineating a subculture in need of preservation, perhaps akin to a threatened ethnic community. Under this perspective, selecting for a deaf child would not be much different than choosing the sex of the baby or the skin color. Dean and Tracey may argue that they are merely fulfilling their right to reproductive autonomy, to make their own choices in the same way that someone using contraception is making a reproductive choice.

As the cases in this chapter demonstrate, the use of IVF, and more specifically of PGD, raise complex questions about which it is difficult to draw up legislation. There are very real issues about where a line between appropriate and inappropriate uses could or should be drawn. Nevertheless, it is important that scientists, ethicists, lawyers, and politicians do not shirk their responsibility to grapple with these difficult questions in order that we avoid a drift into eugenics where those who can afford to do so will be able to exploit new genetic tests to their own advantage.

THE DEBATE

IN FAVOR:

- PGD allows us to break the cycle of inherited diseases.
- It can provide treatment for sick relatives.
- It emphasizes autonomy, our freedom to choose.

AGAINST:

- An important number of embryos are discarded when PGD is performed.
- It can lead to eugenics and targeted selection of physical features based on fashion or culturally relevant characteristics.
- It can also lead to the increase of children born with disabilities (as a side effect of IVF).
- Possible psychological impact on savior siblings.
- Negative perception of people with disabilities or chronic illness.

CHAPTER 2

Haven't I Seen You Before?

HERE WE GO AGAIN

Mandy was eight years old when Rocky died. Right from birth, that little dog had been Mandy's constant companion. They had spent countless hours playing in the backyard and at the park, running and goofing about. And now her best friend was gone. Why did this have to happen to her? Life was so unfair!

Mandy's father had walked out of their family when she was just a newborn; she had no memory whatsoever of him. Once bitten, twice shy—Sue had never sought to bring another father-figure into their lives, electing instead to bring Mandy up alone. As a result, the bond between mother and daughter was unusually strong. It wasn't an easy decision; it is physically and emotionally demanding to be a lone parent, but Sue was fortunate to have financial security courtesy of the family business. She knew that a lot of other women in a similar position were far worse off than she was.

Mandy had always been a contented girl, but with Rocky gone, her personality was dramatically altered. It was tearing Sue apart to see Mandy like this, sobbing in her room . . . Surely there was something she could do to comfort her child!

With a flash of inspiration, Sue recalled an advertisement she'd seen in a Sunday paper a couple of months earlier. It was a half-page, full-color ad. At the time it hadn't seemed directly relevant to their circumstances, so she quickly forgot about it. But there was now real urgency about tracing that company.

Armed with the little information she could remember, Sue turned to the Internet. Luckily, it didn't take her long to get the contact details she needed. The Web site of *Companions Forever* had the same color scheme as the advertisement she'd seen and the details were explained using the same phrases. "The science of cloning is here—death has been defeated! Allow *Companions Forever* to bring your beloved pet back to you."

Instinctively, Sue might have been skeptical. But she'd recently seen a report about pet cloning on that human interest show that comes on after the evening news. To be honest, the scientific details had gone over her head, but there was no doubt that this was real. The scientists had shown **cloned** cats and dogs, and they looked authentic. Whatever it might cost, she knew that she had to bring Rocky back into Mandy's life.

Sue called the number and set an appointment right away. They said that they preferred to prepare clones in advance of a pet's death to provide "uninterrupted companionship" but they were also able to clone a dead pet provided that just a small sample of biological material was available. They recommended a few hairs taken from the pet's grooming brush or basket. Sue was a little worried at this stage: she'd already washed Rocky's basket and a fear that she'd have to dig him up in order to get the necessary DNA crossed her mind. Fortunately, however, his brush came to her rescue. Sizeable quantities of Rocky's coat hair were trapped there.

The following day she headed down to their office, clutching an envelope of dog hair in her hands. She wrote them a check for the necessary amount without thinking twice. No price was too high to bring back happiness to Mandy!

True to their word, the company contacted Sue a few months later to let her know that her clone was ready for collection. Mandy's face lit up when she opened the box; she had no idea what her mother had been up to on her behalf. Having that little white puppy in her arms was the best that ever happened to her. Sue just watched her play with Rocky II and said nothing. Mandy's *joie de vivre* had returned. Sue was proud of what she had done.

Over the next few months, a bond formed between Mandy and Rocky II. It didn't matter that he didn't respond in quite the same ways as the original. His personality was clearly different; he wasn't as warm and playful as the first Rocky had been. He looked almost identical to him at the same age, but Sue and Mandy could both tell that it wasn't the same dog. Neither mentioned it to the other. Deep down, however, it didn't matter—he was close enough, and that worked for everybody.

VOCABULARY

Clone: an identical copy of something. In this context, a genetic copy of an organism.

Epigenetics: scientific discipline that studies how the environment affects the genome without actually changing its original genetic information. Epigenetic changes can influence how genes are activated and thus affect the appearance and physiology of an organism. They can also be transmitted to the next generation under certain circumstances.

Eugenics: scientific and social strategies aimed at improving the genetic composition of a population. Usually applied to the human race, as a way to select and favor "better" men.

As the years went by, Rocky II grew into a family member as loved as the original Rocky had been. Mandy, of course, also grew. As she did so, she developed a circle of friends her own age and she would go out with them more and more. Rocky II seemed to be content to spend time with Sue rather than Mandy. Sue treasured his companionship, curled up on her lap as she sat at home reading and waiting for Mandy to get back from school or the movies or a shopping trip with her friends.

One Saturday afternoon, Mandy didn't reappear at the time they agreed. This was unlike her; she may have become a teenager, but that bond with her mother was still uncharacteristically strong. Sue was starting to panic, and her fears were realized when a call from the local hospital asked her to come there urgently. A tire on a truck had burst, causing it to skid out of control. It had plowed into a group of girls heading home from the mall.

Sue ordered a taxi—she didn't feel safe to drive under the circumstances—and rushed to the hospital. She arrived too late: Mandy had experienced a cardiac arrest and died. The sight of her daughter's lifeless body was more than she could take. Her whole world was crumbling in front of her. Mandy had been her reason for living for the past 13 years. Sue couldn't believe that her precious daughter had gone, lost forever. She just couldn't give up.

Sue asked a nurse for a glass of water and waited until she left the room. Then, she caressed Mandy's blonde hair and with a sharp pull, took a sizeable blonde lock from behind Mandy's ear, taking care that it didn't show.

Having kissed her daughter one more time, she ran away, clutching the hair firmly in her hand.

Five years had passed since Rocky had been cloned. Although no one had officially cloned a human yet, it was common knowledge that many of the earlier difficulties with applying the technique to humans had been resolved. She'd even seen a doctor on the news saying it was now possible. The other guest on the show had been dismissive, openly laughing at him. But what if he was right? He *had* to be right.

She had money . . . and she had Mandy's hair. She would bring her back, the price was not an issue. Even if the new Mandy turned out to be a little different, more like a twin sister than a true copy . . . that was much better than having no Mandy at all.

THINK ABOUT IT . . .

Would you pay to clone a dead pet? Would it be money well spent, or would you be better off simply getting a new one?

What about cloning a family member or loved one? Would this be worth it, even knowing that they wouldn't be an exact copy, just a new person, although physically identical to the deceased?

Would you want to be cloned after dying?

NOT JUST SCIENCE FICTION

Pet cloning services may sound far-fetched, but they really do exist. Dr. Woo-Suk Hwang is a South Korean scientist who fell from grace when it turned out that his breakthrough in producing personalized stem cells for human patients was faked (see Chapter 3). As part of the investigation into his other research, scientists checked whether his claims to have cloned a dog were true. And it turned out that they were. The stem cells may have been bogus, but Afghan hound SNUppy (named after the Seoul National University, where the work was done) was the real deal.[1]

Byeonn-Chun Lee, one of Hwang Woo-Suk's colleagues in the successful dog-cloning project, set up a company offering to replicate your precious canine companion. It's not straightforward and it's not cheap, but for the right money he said they could reproduce your dog (though they would need a better genetic sample than you'd get from a few stray dog hairs). One of the earliest people to use the service was Vernon McKernie, who paid about $100,000 in 2008 to have five back-up copies of her late dog Booger.[2]

HWANG'S PET PROJECT

In 2008, a company called BioArts International was set up with the goal of offering the first pet cloning services to the public. They did so through a partnership with the lab of Korean cloner Hwang Woo-Suk.

They initially auctioned the possibility of cloning any dog, at a starting price of $100,000, a program they called Best Friends Forever. They got five undisclosed offers and in September 2009, they delivered five cloned puppies to their customers. However, the same day they announced that they were suspending the service, citing legal patent issues, economic difficulties, and certain technical problems. According to BioArts's CEO, "Dr. Hwang's technology is not ready for prime time."

If your best friend happens to be a cat, then cloning has been possible for even longer. The first feline clone, called CC (short for CopyCat), was born in December 2001.[3] While some companies are getting in on the act, others have stated that it may be too early to start serially cloning pets for business (see box: Hwang's Pet Project).[4]

So far it seems that cloning humans is a completely different story. The concept of making a copy of a person has been a cornerstone of science fiction for decades (see box: Hollywood Copycats). For the most part, these stories have presented human cloning as evil or as part of a twisted way to enslave humanity. Given this background, it's not surprising that news in 1996 of the creation of Dolly the sheep,[5] the first cloned mammal, was received with a mix of skepticism and fear. Was one of the wildest dreams of science fiction finally possible? And would that mean that we would eventually see all those doomsday scenarios come true?

As it turns out, for technical as well as ethical reasons, human cloning has not followed as swiftly after Dolly as some had anticipated. Even the production of that famous sheep had been a monumental struggle. It took Dr. Ian Wilmut's team at the Roslin Institute in Edinburgh 277 attempts (each requiring a new egg) to achieve nine pregnancies, of which only one ended in a successful birth. All the available evidence suggests that getting the technique to work for humans would be even harder. For this and other reasons, cloning humans remains impossible. Or does it?

MAD DOCTORS AND MAVERICKS

Scientists are often portrayed in fiction as wild-eyed (and wild-haired) egomaniacs prepared to do anything—legal or illegal—to achieve their aims. Fortunately, this image only matches a very small minority of all scientists. In no area is this stereotyping more common than in the field of cloning. There is something peculiarly unsettling about the concept of being able to create people at will, all the more so if they are all mini-Hitlers (as in *The Boys from Brazil*) or a clone army (as in the *Star Wars* saga).

It has to be stressed that most experts insist that reproductive cloning is not an interesting goal for them, beyond the mere technical challenge. The consensus is that there are no true benefits for humankind in making genetic copies of somebody, as we will discuss later. Cloning also raises the specter of **eugenics** (the idea of improving the genetic pool of the human race), since presumably we would only want to clone the "best" (and probably richest) individuals. For many people, this sounds too similar to some of the theories that the Nazis favored in the mid-20th century. That is why nobody seems to be investing time and effort into cracking this problem, at least not publicly (with a couple of exceptions, as we will see). Indeed, at the time of this writing, nobody has presented any proof that human cloning

---

HOLLYWOOD COPYCATS

Even before it was possible to clone a mammal, we have been familiar with the concept thanks to books and movies. Maybe that's part of the problem: the unrealistic portrayal of cloning in science fiction, in which they are usually involved in some evil scheme, has been embedded in our collective minds, generating unreasonable fear.

The earliest depictions of clones as monsters can be seen in classic science fiction horror films like *The Invasion of the Body Snatchers* (1956) and *I Married a Monster from Outer Space* (1958). In the seventies, *The Clones* (1973), *Embryo* (1976) or *The Boys from Brazil* (1978) showed that clones were still able to instill panic, but used more "realistic" plots. Resorting to clones was even more frequent in subsequent decades, often looking for a comedic effect, with blockbusters like *Multiplicity* (1996), *Austin Powers* (1997), *Alien: Resurrection* (1997), *The Sixth Day* (2000), or the *Star Wars* trilogies, in which a full army of clones is used to control a galaxy.

has actually been carried to completion, and according to most experts, this is unlikely to happen any time soon. Not only is human cloning complex, it is also banned in the official science programs of many countries.

Of course, technical difficulties can eventually be overcome, and formal bans can be ignored provided you have an independent source of funding. It is possible that sooner or later someone, somewhere may be able to achieve it . . . and some people are certainly trying. On the fringes of reproductive science, several individuals have claimed both that they are actively pursuing cloning and that they are very close to achieving their goals. The most visible are Panayiotis Zavos and Severino Antinori. Dr. Zavos is an American-Cypriot expert in reproduction, who in the early years of this century announced he was going to attempt to clone a human. His comments sparked enraged cries from the public and scientists alike. Since then, he has stated several times that he has already succeeded in completing the first stages of the procedure (see box: Attack of the Mad Cloner).[6]

Dr. Antinori is an Italian gynecologist who has never been afraid to be seen pushing the technological and ethical boundaries in reproductive medicine. For instance, he has achieved certain notoriety for helping women over 60 to become pregnant. Antinori formed an alliance with Zavos and he has also announced that he has successfully implanted cloned embryos in human wombs.

ATTACK OF THE MAD CLONER

In 2003, Panayiotis Zavos announced that he had created the first human cloned embryo. He did not provide further details or any proof of his claims. Then in 2004, he said that he had implanted a cloned embryo into the womb of a volunteer. He had generated it mixing an egg of the 35-year-old donor and the nucleus of the skin cell of her husband. He was then waiting for the embryo to develop, which he suggested a 30 percent chance of happening. However, it was not successful. Once more, he did not provide any evidence that he had actually achieved these initial steps of cloning. In 2009, he announced that he had obtained 14 cloned embryos and had transferred 11 of them into the wombs of volunteers. This time he recorded a video to document the process of cloning, but still failed to produce scientific evidence to endorse his claims. As of today, there is no news of any successful pregnancy or birth using Zavos's techniques.

Is any of this true? Are we really that close to seeing the first human clone? No evidence has yet been produced to support the claims of Zavos and Antinori. But it may only be a matter of time, and the serious players may be rather less keen to portray their work in the full glare of the media spotlight.

Are the appearances on TV of mavericks such as Zavos and Antinori a mildly amusing distraction, or do they offer insight into a sinister side of science? Should we attempt to control them in some way or other? It can be argued that scientific freedom is one of the things that has allowed society to advance, but does that mean we should allow an "If we can, we should" mentality to dominate? Human cloning could be a serious threat to human dignity, as we will discuss below. Some religions definitely see it this way: Islam and Catholicism have already condemned it strongly.[7] Are all these precautions really needed, or are we unnecessarily interfering with scientific progress?

As is so often the case, the problem is more complicated than it initially seems, and careful thinking is required. In 2005, the United Nations published the *Declaration on Human Cloning*,[8] in which it was stated that reproductive cloning goes against human rights and should be forbidden in all member states. The European Union has already passed laws that endorse this view,[9] while leaving the door open to the kind of therapeutic cloning techniques we will describe in the next chapter. Australia has similar legislation in place.[10] However, as we have already noted, this is no guarantee that human cloning will never happen. If we think that real clones may one day roam the Earth, we need to maneuver past the movie-inspired reactions and start to look at the genuine implications this may have in our society (see box: Blast from the Past).[11]

BRINGING BACK THE DEAD?

We have talked so far about the current state of the cloning techniques, which seem to be far from optimal for such a difficult task. The question that we should be asking ourselves next is: if one day we truly have the tools to clone humans, what could we gain from it? The advocates of reproductive cloning argue that this could help infertile couples to conceive children that would at least be genetically related to one of them. However, other techniques of assisted reproduction could currently achieve this, and other cloning-free advances are constantly being made. Therefore, this seems to offer meager benefit for significant ethical (and financial) cost.

If we push the theory to the limit, we could imagine a scenario in which we could have cloned bodies into which we would transplant our brains.

BLAST FROM THE PAST

So far scientists have successfully cloned several species of animals. Could cloning be used to "revive" species that have been extinct? This was impossible when Michael Crichton wrote *Jurassic Park* in 1990, a novel in which scientists created a zoo full of "resurrected" dinosaurs.

However, the techniques continue to be refined, and there have recently been serious discussions about the possibility of cloning a mammoth. In fact, a first draft of the sequence of its genome has been completed, albeit with notable gaps in the readout. This may not be a problem, however, if the missing sequences can be substituted using elephant DNA (in much the same way that the scientists in *Jurassic Park* used other reptile DNA to overcome shortcomings in their dinosaur sequences). Female elephants would also be expected to act as surrogate mothers for the cloned mammoths. It may be a distant dream but, not for the first time, reality is starting to echo science fiction.

People have also speculated about cloning a Neanderthal, for which a draft genome has also been obtained from fossilized remains. However, experts doubt that we will ever be able to complete this sequence, since the biological material needed to do so is not of the sufficient quality.

This would allow us to replace old organs in a more radical way than that offered by regenerative medicine. Brain transplants would be complex (think of the hundreds of nerves that would have to be patched correctly), and is likely to remain science fiction for a while at least—although some surgeons are starting to speculate about the feasibility of a human head transplant.[12] From another perspective, however, the use of clones as sources of organs for transplants could indeed be a feasible (but highly problematic) possibility (see box: The Clone Farms).

Armies of cloned soldiers may seem an interesting prospect to some, but it involves more than physical cloning to produce hundreds of individuals, each with their will molded to that of their leader. Once again, this seems better suited to fiction than reality.

Other than that, it's difficult to picture reasons why human reproductive cloning could be useful or even interesting. The idea that cloning would be like bringing somebody back from the dead completely ignores the

THE CLONE FARMS

Another scenario inherited from science fiction books and movies is the possibility of creating a clone just to harvest its organs when needed. Having the same genome would prevent any rejection when an organ is transplanted. The films *The Island* (2005) and *Never Let Me Go* (2010) deal with this possibility. These films offer contrasting views of the impact being a walking spare-parts factory has on the clones themselves. In *Never Let Me Go* they seem curiously resigned to their fate, whereas in *The Island* the discovery of their purpose causes them to run . . . which would likely be the most common reaction.

importance of environmental influences on the way we behave and look. Even while you were developing in your mother's uterus, the environment was playing an important role in shaping the person you are, both physically and mentally. Decades of studies have been carried out on identical twins, and while these have shown that a lot of the similarities arise from shared genetics, there are nevertheless a lot of differences, even when we are considering two people who shared the same womb. In the case of clones, of course, we'd be talking about a different womb, at a different time, which would increase these differences even further.

In recent years, a whole new field called **epigenetics** has emerged, which investigates the changes in the way genes within our DNA are switched on or off under the influence of other factors, including our environment.[13] In other words, we are uncovering the molecular mechanisms by which the environment can modify how our genes behave, and how these changes not only affect us, but can also have the potential to affect our descendants. The influence of the environment on our genes can continue throughout our lives, which is one of the reasons why the longer they live, the more identical twins end up diverging.

There's no getting away from it: a clone simply wouldn't be the same person as the original, despite initially sharing the same DNA. Neither would cloning ever be a way to truly achieve immortality. Leaving behind a copy of your unique combination of genes when you die could be a way of somehow prolonging your "impact" on society, but it would not be a continuation of "you." A clone would not only have mental and physical differences, but it would not share any of your memories or past experiences because these are not transferred via your genes.

HUMAN RIGHTS, CLONE RIGHTS

Let's move on to consider things from the perspective of the clone. Learning that you are a genetic copy of somebody is inevitably going to have a psychological impact. First, there is the issue of having been "created" in a lab in order to achieve a "goal," whether that be just to show off a technical ability or for some other selfish objective (like Sue's intention to replace her dead daughter in our story). This would lead to the pressure to achieve the purposes for which you were made. This issue was considered in the German film *Blueprint* (2003), in which a pianist suffering from a progressive neurological disorder has herself cloned so that her daughter can take over performances.

From a religious perspective, cloning humans has been seen as stepping into divine territory, almost akin to creating life from scratch. It goes beyond the normal techniques of assisted reproduction and creates several further conundrums. For example, consider the notion of a spiritual essence or "soul," a concept found in many religions. Would a soul be shared with our clone, or would it be different? It's more than just a metaphysical question because, depending on the answer, clones could either be considered fully human by society or just as some kind of soulless lower entity. There are echoes here of existing debates about the "personhood" of the extremely handicapped and, from our not-so-distant history, about people considered "inferior" for other reasons, such as the color of their skin. Would these be resuscitated if human cloning becomes a reality?

Moreover, the relationship with our own clone would be difficult to classify. There would be an inevitable age difference between the original and the replica. How would this affect their interpersonal dynamic? Would they feel more like twin brothers or like father and son? We may get a hint of this in the near future. As we have said, with the advent of IVF and the possibility of embryo freezing, there are now biological twins, conceived at the same time, but born many years apart.[14] Any psychological issues they manifest coming to terms with the age difference will give us clues about the much greater issues faced by a clone.

WOULD HUMAN CLONES BE NORMAL?

We have described some of the social and psychological impacts of being a clone. We also have to consider the physical consequences. After all, a clone would start from DNA taken from the nucleus of an adult cell. During the lifetime of the original person, DNA within that cell would inevitably have been subjected to an unknown amount of damage associated with

SO, YOU'RE MY CLONE, THEN?

One reason for forbidding human cloning is the potential to clone someone without their knowledge, e.g., using genes taken from a biological sample they'd left somewhere, perhaps when they cut their hand or even when they combed their hair. You might think that there's no risk that anyone would bother to duplicate you, but the temptation to "copy" a top footballer or a rock star might prove overwhelming. Also, this could be seen as forcing somebody to reproduce against his will (or "stolen parenthood," as it has been called), something that contradicts the most basic human rights.

the aging process. Would that affect the life of a clone? Would clones be as healthy as normal individuals?

We don't have a definitive answer to these questions. Even in the cases of nonhuman cloning that have already occurred, interpretation of the evidence is not straightforward. Dolly the sheep certainly died at the premature age of six, suffering from a rare cancer. It wasn't clear that this was due to the fact that her original genome was not as "fresh" as the one that would have been passed by a sperm and an egg or was simply a coincidence. Since then, other cloned animals have been carefully monitored during life and studied after death. Some of them have presented all sorts of anomalies and health problems, but others appeared to be completely normal, even being able to reproduce and live to an old age. BioArts International cited "occasional physical anomalies" in their cloned dogs and "unpredictable results" as reasons why they believed the current cloning techniques were not ready.[15] These may be hints of problems to come in the field of cloning (and opponents have certainly presented it as such), but until we have enough evidence to see statistically significant patterns emerging, we will have to withhold judgement.

In the meantime, if there is even the possibility of such problems, we ought only to proceed with extreme caution. We will undoubtedly have to have a much clearer picture before embarking on human cloning. Inefficiencies in the cloning process, the risk of producing a defective embryo or an individual that will die prematurely, need to be seriously assessed. No government should be willing to give the green light to cloning while such uncertainties remain. We still have a lot to learn before being able to safely apply the technique to ourselves, and hopefully awareness of these limitations will be a deterrent to anyone considering jumping into the cloning arena without careful reflection on the possible ethical consequences.

THE DEBATE

IN FAVOR:

- It would help couples that can't have children by any other method.
- It would be a major technologic advance.

AGAINST:

- Potential of creating second-class citizens and human rights abuse.
- No clear benefits to humankind.
- Possible physical and psychological harm to clones.

CHAPTER 3

Exchange Parts for Everybody

THE END OF THE ROAD

Mr. Wilson tries to catch his breath. Climbing that last flight of stairs really got the better of him. Dr. Ganesh has his office on the top floor of the clinic, and the elevator is not working. "Just my luck," thinks Mr. Wilson, as he staggers into the waiting area. He's going to feel his knee for the rest of the week, that's for sure.

The nurse asks him to sit and gives him a glass of water. He just nods and follows her instructions. He is there a few minutes before Dr. Ganesh comes out to greet him. By the time he does so, Mr. Wilson is just about capable of conversation, but his throbbing knee lets him know that full recovery will take several days. He hopes the journey has been worth it and that these problems will shortly be over.

"I'm sorry that the lift is out of order today. Are you feeling OK?" asks Dr. Ganesh.

In reality, there's no need for Mr. Wilson to reply. It's obvious that he is not in the greatest shape. Dr. Ganesh indicates towards the consultation room, and Mr. Wilson hobbles in.

With the door barely closed, and with Mr. Wilson still flopping into his chair, Dr. Ganesh continues.

"We have the results of the tests we performed last week," he offers. Then looking up from his notes to ensure eye contact with Mr. Wilson he adds: "I'm afraid I don't have good news . . ."

Mr. Wilson hates it when doctors adopt a serious face and say things like that. They both know the results are not a surprise to anyone. He just wishes that they would bypass these useless introductions and get on with the details.

"Just tell me the extent of the damage, please, Doctor."

Dr. Ganesh pulls his file up onto the computer screen.

"Let's begin with your lungs. They're in pretty bad shape: you have already lost 80 percent of capacity, and without intervention I think they are likely to fail entirely within the next 12 months. Your right knee, as you demonstrated on the way in, is showing increased problems. Realistically, you are going to need surgery in the very near future."

"How about the left one?" interrupts Mr. Wilson.

"Ah yes, I was coming to that. The left knee is not quite as far gone. However, without treatment I feel you are likely to lose the ability to walk entirely within the next couple of years."

"Is that all? This doesn't sound so bad," says Mr. Wilson, with more than a hint of irony.

"I'm afraid there is more," continues Dr. Ganesh after double-checking the computer screen. "Your eyesight has shown a 40 percent deterioration since your previous checkup."

"I noticed."

"And finally, your liver is giving up. The serological values are through the roof, some of the worst I've seen during my entire career, to tell you the truth . . ."

Mr. Wilson can't repress a smirk. He's not surprised at all. Old habits die hard. And he has a few of them. He never refused a drink and he's been a heavy smoker for the last 50 years. He never cared that much about eating healthy foods or exercising either, that's a fact. It's no wonder that his body is in such dire condition. It's actually surprising that he's survived this long.

Dr. Ganesh is thinking exactly the same, but avoids saying so. That would be pointless. He keeps his serious face. He's done this hundreds of times and he knows it's important to make sure that the patient understands how bad his illnesses are.

Mr. Wilson is no fool. He's worked in sales and he knows that Ganesh is building towards the big pitch. But he also knew what he was getting into when he contacted the clinic, and he's not going to let Ganesh take the upper hand.

"OK, let's cut to the chase. How much is this going to cost me?"

VOCABULARY

Embryonic cell lines: cells obtained from human embryos and cultured in the lab for long periods of time.

Embryonic stem cells: "master" cells that are truly pluripotent and might be more suitable for the requirements of regenerative medicine.

iPS cells: induced pluripotent stem cells. Cells from an adult organism (which are called *somatic cells*) that are genetically or chemically reprogrammed to behave like stem cells.

Oocytes (or **"eggs"**): cells produced by the ovaries that, when fertilized, start dividing and form an embryo.

Pluripotent: those stem cells that can, in principle, generate any kind of cell. Embryonic stem cells are usually pluripotent. Other kinds of stem cells, like those found in adult tissues, are usually not.

Regenerative medicine: new medical studies aimed at substituting tissue and organs with new ones created in the lab, in order to cure diseases.

Reproductive cloning: to generate an embryo with the DNA of an adult, in which the goal is to create an adult genetically identical to the donor.

Somatic cell nuclear transfer (SCNT): to transfer the nucleus of an adult cell into an oocyte that had its own nucleus removed. A technique used in cloning.

Stem cells: "master" cells capable of generating all (or some) of the different types of specialist cells within an organism. Most cells within an early embryo possess this potential, hence the interest in embryonic stem cells.

Therapeutic cloning: to generate an embryo with the DNA of an adult, with the sole purpose of obtaining personalized stem cells for regenerative medicine treatments.

A broad smile fills Dr. Ganesh's face. He presses a button and a white screen descends from the ceiling. A projector immediately starts spitting pictures of lungs, hearts, eyes, livers. Healthy looking organs, clean and ready to use.

"Our tissues are top of the line, as you can imagine. Grown in our own labs from the finest **stem cells** available. They can be ready in just a few weeks. We have your DNA stored already, so we could transfer it to the stem cells right away. We should have the transplants available before Christmas. And by New Year you will indeed be a new man."

A list of numbers takes over the pictures on the screen. Mr. Wilson looks at them for a second, then nods. He can afford that. It's in fact quite a low price to pay to be able to cheat death and go on living in a better, healthier, and younger body. Until it's time to replace the damaged pieces one more time, of course.

"Sounds good, Doctor," he says. "Actually, now that I think about it, I've noticed that my hearing is not as good as it was . . ."

Dr. Ganesh types something into his computer.

"Of course. We will take care of that too. I'm afraid there is a 25 percent deposit up front. I'm sure you are aware that the process of cell culturing is not cheap . . ."

"Sure, I understand," replies Mr. Wilson.

He passes his credit card to the doctor. After the transaction is done, they shake hands and Mr. Wilson leaves the office. As soon as he's passed through the reception area he takes a cigar from his breast pocket and lights it before starting a slow descent of the stairs.

THINK ABOUT IT . . .

Should people be required to show they have taken good care of their original bodies before having access to replacement organs?

Is it wrong to pour money into development of organ replacement treatment that only the rich can afford while millions of people suffer with diseases that might be cured for a fraction of the cost?

Would you want to replace your organs if to do so required the destruction of human embryos?

PART EXCHANGE

When a part of our body is about to stop working, our best option nowadays is to substitute it with another before it's too late. The more critical the organ, the more urgent is the need to have a transplant. At present, the new tissue needs to be obtained from someone else: a donor.

There are numerous problems associated with donations. Organs cannot be stored for long periods of time: they need to be passed to the recipient within a matter of a few hours at most. In situations where we have two copies (the kidneys, for example), one can be given by a living donor. But in other circumstances, the donor must be reasonably young and have recently died in order for the tissue to be available and of good enough quality. There may therefore be a race to get the organ to the recipient in time.

There are even more pressing problems. The most important is probably that supply of organs cannot keep pace with demand. Even in countries where consent to give your organs at death is presumed, there are still people dying on the waiting list because a suitable organ doesn't become available in time (and there are plenty of countries where organ donation is voluntary; see box: Opt In, Opt Out).[1]

There's another major obstacle that prevents transplantation from being a long-term solution to a serious health problem. Our immune system is finely tuned to fight off foreign material, such as bacteria and viruses. Obviously, taking a lump of tissue from a different person and plumbing it into our body's vital systems sends our immune defenses into overdrive. Without a series of safeguards, the organ would rapidly be rejected and destroyed. We can mitigate some of these difficulties by choosing donors who

OPT IN, OPT OUT

In some countries, everybody is considered a potential donor unless they have expressly stated that they do not want to be. This system is called *opt-out*. The opposite model is *opt-in*, in which one has to voluntarily register as a donor. Naturally, countries with opt-out systems (like Spain, Sweden, or Austria, among a total of 24 European countries) tend to have higher donor rates than those that use the opt-in (such as Germany, Greece, or the United Kingdom).

Spain is one of the countries with the highest rate of donors (34 for every million habitants). It also has the hardest opt-out system, but that is not the only explanation for the success. The establishment of a network in hospitals, better coordination, and a greater sense of social responsibility through public campaigns have significantly contributed to the current situation.

are genetically similar to the patient (relatives, for instance) and by long-term use of immunosuppressants, drugs that basically shut down our defenses. This is the best way to ensure that the transplant is accepted, but there is a clear drawback: it also leaves us open to attack by external agents, since we can no longer effectively fight the microbes that invade us. The consequence is that people who have received a transplanted organ are therefore more susceptible to serious infections. And even with the best combination of genetic match and drugs, there is still a high risk of early rejection, sometimes for reasons that we don't fully understand. Moreover, most transplanted organs fail after a number of years (often between 10 and 20, depending on several factors),[2] likely because it is impossible to completely prevent the recipient's immune response against it.

Alternative treatments are sometimes feasible. Dialysis, for instance, is a procedure by which a machine performs the functions that the kidneys would normally do, namely "clean" the blood of toxic substances. However, most dialysis machines are not portable; dialysis requires the patient to go to the place where the machine is located and to stay there for several hours while their blood passes through a series of filters. This process is exhausting and needs to be repeated every few days, so this is far from being an optimal solution.

Other illnesses can be overcome in simpler ways. For example, the pancreas normally makes insulin, a substance we need to process the sugars that we eat. In some forms of diabetes, the levels of insulin are not high enough, normally because the cells of the pancreas are not working properly. This leads to a dangerous increase in blood sugar, which can, in turn, cause serious complications, including death. If the diabetes is mild, there are certain pills that can stimulate a "lazy" pancreas to make more insulin.[3] But if the pancreas can't work any harder, the only option is to inject the missing insulin (you can't take it as a pill as the body would mistake it for food and break it down). Millions of diabetics give themselves daily injections. The process is inconvenient and sometimes problematic, but at least it doesn't require an organ replacement or bulky and expensive machines.

These different options (transplantations, mechanical substitutes, and drugs) have been responsible for saving many lives all over the world in the past decades. But, as we just discussed, they are not sufficient to solve all the problems related to organ failure, especially those associated with normal aging. It is at this point that we encounter the potential for **regenerative medicine**. The principle is simple: grow tissues in the lab to replace those that don't work with the goal of creating your own genetically matched organs for transplant, and in so doing restore all lost functions. In theory, this could also be extended to milder conditions that do not involve total organ failure. Rather than tackling the symptoms of a disease, it might be possible to correct the underlying cause of the problem, freeing up the patient from frequent and painful medication and possibly saving their life.

Regenerative medicine sounds like something out of a futuristic movie, but it could become the single most important advance in the recent history of medicine . . . if, of course, we can ever get it to work. Real breakthroughs are starting to take place, but there are still plenty of technical kinks to iron out. Some of the difficulties may never be resolved—who knows—but in the meantime, this new approach is sparking genuine excitement. And genuine controversy.

STEM CELLS: THE ETHICAL MINEFIELD

If any lab-made organs for patients such as Mr. Wilson are going to be produced, then *stem cells* are likely to have a crucial role to play in the process. Right now they are the basic ingredients for growing fresh tissue of any kind. By definition, these are capable of giving rise to a wide variety of cells within the body. Exactly how varied depends on the type of stem cells since, as we will see, not all are the same.

Where can we get these precious cells? In our journey into existence, we start as a single cell (an egg, from the mother, fertilized by a spermatozoid, from the father). That single fertilized cell divides into two cells, then again to become four, and so on. Eventually, we end up with a body consisting of billions of cells, the majority of which have become specialized into very different types, such as brain cells, skin cells, or bone cells. It stands to reason, therefore, that early in this process there must be some "universal" cells that can turn into every type of cell in the body. These initial cells are essentially "blanks" that can evolve into anything the body needs. This phenomenon is called *pluripotency*.

The ability to isolate these **pluripotent** stem cells is seen as a crucial first step in regenerative medicine. Once we have such cells, we would then need to find a way to push them into the direction we want in order to generate particular organs in the lab. Finally, we would need to transplant the new organ into the recipient. The first problem, however, is getting hold of pluripotent stem cells. There are more than one potential source, and all have pros and cons.

The existence of stem cells was proposed in the early 20th century, but it wasn't until the 1960s that the theory could actually be proven. At that stage, stem cells were found in adults, initially in bone marrow. These cells could divide and form any kind of blood cell required, but could not naturally be turned into other sorts of cells.

A LEGAL LOOPHOLE

The United States has become one of the most important battlefields in matters of stem cell ethics. This is probably due to the combination of being one of the countries with the most advanced biomedical research network as well as having a big percentage of the population with strong religious views.

During the George W. Bush years (2000–2008), public funds to study stem cells were limited. This position was reversed by Barack Obama in March 2009. But in the summer 2010, pro-life associations found a way to freeze all research done in embryonic stem cells.

Two scientists, James Sherley and Theresa Deisher, backed by anti-stem-cell groups, argued that transferring funding to embryonic stem cells work was damaging their own research on other kinds of stem cells, because they had to compete for the same pot of money. This legal loophole was sufficient to temporarily stop all research done on federally funded stem cells. In September 2010, after complaints from a number of other prominent scientists, the research was allowed to continue while its legality was being studied by the Court of Appeals.

In spring 2011, the judges ruled that the research should indeed continue, a decision that was upheld in a decision taken the following July.

Another breakthrough came in the 1980s, when experiments using mice helped scientists uncover **embryonic stem cells**.[4] These cells are truly pluripotent and may be more suitable for the requirements of regenerative medicine. In the 1990s finally it became possible to obtain cells from human embryos and culture them in the lab for long periods of time (the so-called **embryonic cell lines**),[5] heightening both the excitement and the controversy (see box: A Legal Loophole).[6]

In view of this, obtaining pluripotent stem cells for regenerative medicine raises some big concerns. First, we would need human embryos from which to extract them. Getting "unwanted" embryos is definitely not easy, and animal cells won't do for medicine because they are just too genetically distinct from us. The second and more important problem is that the process of extracting stem cells destroys the embryo. Those who believe that life begins at fertilization see the embryos as "alive" and the process of harvesting stem cells as "killing" the embryo. It's a similar argument to that used to condemn abortion or assisted reproduction on moral or religious grounds. On the other side of the debate, there are those who believe that embryos are not fully human, certainly not while they remain in a glass dish rather than a womb (and possibly even for many weeks beyond that). After all, they argue, if abortion is allowable for later pregnancies then surely the potential benefits of harvesting these cells supersede the "rights" that very early embryos may have.

The starting materials for research have often been "spare" embryos, that is to say embryos produced during fertility treatment but no longer required by their genetic parents, perhaps because they have been successful in achieving pregnancy. The leftover embryos cannot be stored forever. If they are destined to eventually be defrosted and thrown away, how is it not better to make good use of their potential? Opponents would still argue that this line of reasoning puts too much emphasis on embryos as a means to an end and does not take due consideration of the detrimental effect on human dignity.

The sharp contrast between these two points of view has led to heated debates regarding the appropriateness of using embryos as a source of stem cells. Aware of ethical concerns, many governments initially issued moratoria on the use of embryonic stem cells.[7] However, some countries such as Japan and the United States have recently started to relax the regulations to allow further advances in the field, having realized the immense curative potential of this field of research. Others, such as France, renewed their prohibition to use human embryonic stem cells for research.

WHAT ALTERNATIVES DO WE HAVE?

Embryos are not the only place where stem cells can be found. As it turns out, most adult tissues have some, quietly hidden away in a corner, ready to spring back into action. Of these, the easiest to locate are the blood stem cells, which, as we said before, live in the bone marrow. If a tissue needs more cells in any given moment, maybe as a result of an injury that has destroyed part of it, adult stem cells will start dividing and generating the specialized cell required. For instance, in anemia, new red blood cells will be made from the stem cells in the bone marrow.

One thing to keep in mind is that adult stem cells are not as variable as the embryonic ones; they usually give rise only to a limited number of cell types and are said in the jargon to be "multipotent" not "pluripotent." Nevertheless, that would be sufficient in some situations. For instance, liver stem cells would give rise just to cell types common in the liver. Although they would not be useful to turn into neurons, they could very well be sufficient to help regenerate an ailing liver. Using these adult stem cells would greatly reduce the need for the pluripotent embryonic cells. If we could find sources of these tissue-specific cells, extract some without damaging the neighboring tissues, and then grow some new tissue in the lab, we could solve many of the ethical problems.

For a variety of reasons, however, this remains a very big "if." Adult stem cells are rare, difficult to recognize, and usually very hard to separate from the other regular tissue cells. Moreover, most of the systems we have to recognize them also destroy them (stem cells from the bone marrow are an exception, as we will discuss later). Techniques need to improve before we can think of using them for regenerative medicine. Luckily, we are slowly

---

A SELFISH BANK?

Thousands of parents over the world are currently storing their children's cord blood in stem cell banks hoping that one day in the future, regenerative medicine will be advanced enough to take full advantage of these personalized stem cells.

At present, stem cells from an umbilical cord can only be used to treat a handful of rare diseases. In these cases, some argue they don't even have to be a perfect match, which means that genetically close donors would be sufficient.

The first time umbilical cord stem cells were utilized in a transplant was in 1988, and they have proven to be useful and safe. There are currently trials underway studying the use of cord stem cells to treat neurological disorders or diabetes, but it's still too early to know the results.

It has been argued that storing these cells in private banks is wasteful, since making them accessible to everybody by putting them in public banks could mean that certain patients with those rare diseases we mentioned could benefit from them right now. In countries like Spain, private banks have just recently been allowed, and only with the condition that a part of the blood stored has to be publicly available, which has scared off most companies.

Should we be selfish, saving our cells for a magic cure that may never come true, or offer them to save lives here and now, even if it's just for very rare diseases? And, as parents, can we afford not storing our kids' cells, just in case?

starting to overcome these problems. In 2003, the stem cells of the heart were found and isolated,[8] the first step for a potential source of cells to regenerate damaged hearts. The adult stem cells of the lung and colon were not located and identified until 2011.[9]

Other ways to obtain stem cells without destroying embryos are also becoming feasible. For example, there are a good number of stem cells within the umbilical cord. A group of scientists led by Dr. Ispizúa Belmonte at the Center of Regenerative Medicine in Barcelona showed in 2009 that these cells can behave like true embryonic stem cells.[10] Because of this potential, recent years have seen a growth in services offering to collect and store umbilical stem cells for future use (see box: A Selfish Bank?). It is still not known, however, whether stem cells will be recoverable after being frozen for years or even decades. In the meantime, for some it may represent a worthwhile insurance policy to "bank" umbilical stem cells "just in case." The properties of these cells would be much closer to embryonic stem cells than to adult ones and would have the advantage of containing the future patient's own DNA. In the next section we will see why that is so important.

NEXT STEP: PERSONALIZING THE STEM CELLS

We have to remember that the principal issue in any kind of transplant is the rejection of organs because our immune system sees them as "foreign" and sets about destroying them. Using stem cells would not get us around this problem. Unless the stem cells had our own DNA and could be recognized as "one of us," they would also trigger the immune alarms. One way of having "personalized" cells would be to get them from our umbilical cord blood, as we just mentioned. But if we are not lucky enough to have our own stem cells locked away in the bank, what are our options?

For a while it seemed that the solution was going to be what scientists called **somatic cell nuclear transfer**: getting the nucleus (which contains the DNA) of an adult (or *somatic*) cell and putting it into a fertilized egg that previously had its own DNA removed.[11] This egg would then be stimulated to start dividing and form an embryo, from which the stem cells would be harvested.

There are two important issues here. First of all, this is technically very challenging. Although it has been shown to be possible in some animals, it is notoriously difficult to force human eggs to take up foreign DNA. Lack of eggs for the experiments has also been an important limiting point, since extracting them from a female donor (who would have to be first subjected to an intense hormone treatment to stimulate ovulation), is tedious and costly.

But, just like before, there's an even greater ethical concern. This procedure is, in fact, the first step of **reproductive cloning**. It is the technique that was used in the mid-1990s to obtain Dolly the sheep, the first cloned mammal.[12] The fear is, therefore, that success in transferring DNA into eggs for the final goal of making stem cells could be abused to allow full cloning of humans. To differentiate between cloning with the goal of achieving a genetic adult copy of an organism and that made just with the purpose of obtaining personalized stem cells, this latter procedure was branded **therapeutic cloning**.

At the beginning of this century, it seemed that this transfer of DNA into human cells had been achieved by Korean scientist Dr. Hwang Woo-Suk. Unfortunately, this turned out to be a fraud (see box: Hwang, the Fraudster).[13] So far, therefore, it seems that cloning human cells is too technically complicated to use for regenerative medicine. Fortunately a second route that is both easier and less ethically problematic has recently emerged.

HWANG, THE FRAUDSTER

Hwang Woo-Suk was a leading Korean expert in the field of cloning. He was, for example, the first person to successfully clone a dog.

In 2004, he announced that his team were the first to transfer a different person's DNA into a human egg, and eventually create a personalized stem cell. In 2005, he published another paper saying that he did the same thing 11 more times, using DNA from different people and requiring only 185 oocytes, a number much lower than any of the experiments with animals would anticipate.

Later that year, it was discovered that all his personalized stem cells were fake and that he had obtained the eggs illegally (paying for them, which is forbidden in most countries, or requiring female scientists from his lab to donate them). He was found guilty of embezzlement since he had in effect not done the work for which he had been paid. Hwang was suspended but escaped without a prison sentence.

However, he resurfaced in 2008, working for a company in California that cloned pets and attempted to clone pigs in Korea. In 2011, he was trying to secure funds for his research in Libya, where he would collaborate with a local biotech company, when political unrest led to an uprising and he had to be promptly evacuated from the country.

NOW WHAT?

The Hwang affair almost brought the field of regenerative medicine to a stop. All the excitement of being able to infuse our DNA into **oocytes** gave way to the realization that cloning human embryos for therapy might not actually be possible. All of a sudden the production of personalized stem cell therapies seemed completely out of reach.

The pessimistic air was punctured in a dramatic and unexpected way, however. In 2006, Japanese scientist Shinya Yamanaka discovered that it was actually surprisingly simple to take a normal adult cell and turn it back into something that looked a lot like a stem cell. He called these **induced pluripotent stem cells** (iPS cells).[14] Using a skin cell (because they are on the surface of the body and hence easy to obtain) he found that switching on just a handful of genes made it possible to send a specialized cell back to an "earlier stage" of development, a time before it became committed to being the cell type that it was now. A little while later, he and others showed that these iPS cells could become all different sorts of specialist cells. It's a bit like driving off in one direction, deciding that you've turned the wrong way and reversing back into your driveway before heading off in a different route. In theory, by doing this you could turn a simple skin cell into any sort of cell that the patient needed: a neuron, a liver cell, a hair cell, even sperm (see box: iPS Cells and a New Set of Controversies).[15]

This discovery brought about a revolution in the field. All of a sudden it seemed realistic to get a ready supply of stem-like cells without many of the problems associated with the use of embryos. For example, there would be no shortage of starting materials, since we can use any of the billions of cells we have in our bodies. Although this technique typically only works one time in every 10,000 attempts to reprogram a cell, we have plenty to spare. And more importantly, we would not need to transfer DNA into them, since these cells would already have our genetic information.

Not for the first time in the stem cell field, these exciting breakthroughs have been oversold in some parts of the media. The original iPS cell method was certainly not without its own problems—one of the original "cocktails" of genes used by the Yamanaka lab to reprogram the cells is known to have strong links to cancer. Over the last five years, however, the approach of reprogramming adult cells has started to look more and more like the real deal. Scientists are now using a much more streamlined version, omitting the controversial cancer-causing gene, which is not as efficient but it is potentially safer. It has also been observed that cells from fat

iPS CELLS AND A NEW SET OF CONTROVERSIES

In theory, once a skin or fat cell is turned into an iPS cell, it can give rise to any cell type including spermatocytes or oocytes, the so-called *germ cells*. This could be very useful to treat certain types of infertility, since it would help create germ cells for those who have a low sperm count or problems with producing eggs. The important thing is that these cells would have the DNA of the patient, which will in turn be passed on to the following generation. The successful first steps of this technique were published by a group from UCLA in early 2009, and in the summer of 2011, a group from Kyoto University managed to turn embryonic stem cells into sperm and obtained viable pups from them. However, there are still a lot of hurdles to overcome.

There is a potential twist in this story. In theory there's nothing to prevent us from taking skin cells from a woman and turning them into sperm cells. Thus, two females could have a child for which they are both the genuine biological parents. It would also be possible for a gay man to make an egg from one of his cells so that he and his partner could have their own child (though they would have the added complication of needing to rent space in someone else's womb). Although current techniques do not yet allow this, the possibility exists that in the future this will become a reality . . . and another source of controversy.

tissue are easier to induce than those from skin (see box: From Cure to Little Helpers).[16]

The early experiments in iPS cell technology also needed to use viruses to deliver the key genes into cells. Although the viruses had been treated to stop them carrying any harmful genes, there was still the danger that they would cause damage to the recipient cell, leading to cancer or other illnesses. Since then, a variety of scientists have been working to develop virus-free methods of switching on the important genes. Some recent experiments show that simply using selected chemicals may be sufficient to drive the process. In 2010, this gene-free approach was used to convert mouse skin cells into iPS cells and then on to become neurons or heart muscle.[17] Other studies are now reporting the ability to transform human skin cells directly into nerve cells[18] and blood cells,[19] without having to go through the iPS cell stage, which would save time and effort.

FROM CURE TO LITTLE HELPERS

Even if we can't ever use iPS as treatments, they could help us find a cure for many diseases. Take schizophrenia, for example. We still don't understand why it happens. One of the problems is that we can't study the neurons of schizophrenic patients: removing them from the brain through a biopsy would cause more damage that any potential benefit we could get out of them. The iPS could help us bypass this. In May 2011, scientists from the Salk Institute in California got skin cells from a person that suffered schizophrenia, forced them to become iPS, and then obtained neurons out of them. A similar process was done in a patient suffering Parkinson's disease.

At the time of this writing, there remain some issues about whether iPS cells are truly as deprogrammed as it was hoped. Several recent studies reported that mouse iPS cells retain some telltale genetic "markers" revealing their origins, something that would not be expected with an embryonic stem cell.[20] That also seems to be the case in human cells. Even worse could be the fact that iPS cells tend to accumulate unusual genetic modifications in their genomes and their mitochondria, the energy-producing "factories" of the cell. It is not clear how all this could affect their functions and behavior.

Concern has also been shown about another recent study in which iPS cells transplanted into a mouse were attacked by its immune system, despite the fact that the iPS cells were developed from the very same mouse, and therefore had the same DNA. In other words, using your own cells to make iPS cells may not bypass the rejection problems as we expected. If this result is repeated in other experiments, it could represent a new serious inconvenience for regenerative medicine.

Some commentators have questioned the potential of iPS cell technology on other grounds. On the one hand, there are issues of justice. The costs involved in the process of reprogramming personalized cells will put this approach beyond the reach of most patients, leaving it as a treatment only for those who can afford it. Also, there would also need to be substantial changes in the legislation regarding production of iPS cells since current regulations in most countries require separate approval each and every time a new cell line is developed. The paperwork needed for each iPS cell therapy would just be impractical.

SLOWLY GETTING THERE

As we have shown in this chapter, regenerative medicine based on stem cells or iPS cells may have a promising future, provided remaining obstacles can be overcome. Nevertheless, we are still far away from the types of transplant being offered to Mr. Wilson in our initial scenario. There are many issues to be resolved before such therapies become widespread. That said, however, there have been some experimental successes, and clinical trials are currently being performed.

In 2008, a revolutionary trachea transplant was carried out at Barcelona's Hospital Clínic by the team of Dr. Paolo Macchiarini.[21] The trachea, the windpipe going down to the lungs, was taken from a dead donor. If this had been simply cut and spliced into the patient, she would have had major issues with rejection. However, in this case all of the cells lining the trachea had been removed, leaving only the cartilaginous scaffold, which does not in itself trigger an immune response. This was then seeded with cells taken from the patient's own bone marrow, which were grown in the lab for several days. The result was that no immunosuppressants were needed, and there was no rejection. A similar technique was applied to a young Irish boy in 2010, also with complete success. And in July 2011, Dr. Macchiarini went one step further and transplanted what can be considered the first fully synthetic organ. It was also a trachea, but this time the scaffold upon which the patient's stem cells were seeded was not taken from a cadaver but made from scratch in the lab using synthetic

---

THE FIRST LAB-GROWN ORGANS

Between 2004 and 2007, five boys received transplants of lab-grown urethras. These interventions did not use any sort of stem cells. Muscle cells from the bladders of the patients were removed and grown in the lab. Then, they were placed outside a scaffold with the tubular shape and size of the urethra of each boy. After a few days, the scaffolds were totally covered by cells, and then they were surgically implanted in the area where the original urethras were damaged. The lab-grown urethras were still working up to seven years after transplantation. These experiments had a precedent at the end of the 20th century, when bladders grown from muscle cells seeded around a scaffold were successfully transplanted in nine children.

materials (called "nanocomposites").[22] This could be the first step in a series of donor-free transplants, which would be the ultimate goal of regenerative medicine.

Although these are not the only or even the first examples of lab-grown organs (see box: The First Lab-Grown Organs),[23] no one has managed yet to produce a complex one, such as a heart or a liver, which have a range of different cell types in them and an intricate 3D structure. So far we have had striking successes only with tubule-like organs such as windpipes, urethras, or even blood vessels. Functional specialized cells have been made from stem cells and studied in petri dishes, but it is still a significant step to turn these into complex, three-dimensional tissues. Some progress has been made into building an artificial heart starting from a scaffold, but we are nowhere near being able to transplant one.

Research using animals promises some breakthroughs. In 2013, liver tissue grown in the lab was implanted and shown to be working.[24] The same year, lab-grown lung tissue was transplanted into rats and was shown to be able to take up oxygen. And in April 2011, scientists managed to force a group of embryonic stem cells to develop into something that resembled the first stages of the mouse eye.[25] The days when we can turn up to a clinic such as the one visited by Mr. Wilson remain some way off, but the first steps have already been made.

OTHER USES OF STEM CELLS—FACT OR FICTION?

Transplantation of lab-grown tissue is not the only possible avenue for regenerative medicine. Suppose that instead of the need to make a full

---

THE FINAL HURDLE

So far we've been discussing the limits of regenerative medicine due to the difficulties of obtaining the starting materials, namely, a "blank" cell that can turn into another cell type. But there's another pressing issue to resolve. In most of the cases, we still don't know how to force a stem cell (or an iPS cell) to become the kind of cell we want. This is usually done using a cocktail of chemical factors, but we only know the working recipe for a few. For example, we still haven't been able to convince a stem cell to become a beta cell, the pancreas cell that makes insulin, which could be a solution for diabetes.

organ, it proved possible to revitalize an existing one either by injecting our own patient-specific cells or by stimulating the body's own stem cells to take on repair roles. For instance, if our pancreas has stopped making insulin, instead of replacing the whole thing we could inject a few million insulin-making cells to kick-start the faulty organ. This might also prove effective in spinal injuries, where electrical signals between the brain and the extremities of the body have been severed. A localized injection of new neurons could, in theory, "bridge the gap" and cure a paraplegic patient.

Inevitably there are some potential problems with this approach. Firstly, inserting cells in the right place is not always easy. Several clinical trials around the world are attempting this sort of stem cell injection. Most, however, are not using any kind of personalized cells, but just stem cells extracted from embryos. After many false starts, the first clinical trial of an embryonic stem cell therapy, for spinal cord injuries, began in the United States in 2010. It took Geron, the company involved, more than a decade to finish the preparatory experiments and convince the authorities that the procedure was safe enough to test.

At the University of Glasgow, Scotland, scientists are currently injecting fetal stem cells to try to improve brain function in patients after a stroke.[26] There have also been very positive results in a trial in which certain kinds of blindness that arise from damage to the cornea were treated by extracting stem cells from the healthy eye of the patient. In the United Kingdom,

---

BUT IT ALREADY WORKS!

Transplants of adult stem cells have actually been taking place since the 1960s. When someone has a "bone marrow transplant" you are actually transferring stem cells to the recipient. This can be used to treat diseases such as blood cancers (e.g., leukemia and lymphoma), anemia, and immunodeficiencies. They can be obtained from genetically close donors (like brothers) or directly from the patient (this is called an *autologous* transplant). These techniques have been carefully studied and tested, and their efficacy and safety has satisfactorily been proven. However, these results cannot be generalized to other treatments with stem cells. As we have said, blood stem cells are not as pluripotent as embryonic stem cells; therefore it is unlikely that they will be appropriate for most of the regenerative medicine therapies.

a 38-year-old man who had his cornea damaged with ammonia while trying to stop a brawl reportedly recovered his vision via this procedure.[27]

Many other trials are underway or getting ready to start, using adult or embryonic cells, to treat diseases like retinal degeneration. Also, the first studies using iPS cells are also starting. In the next years we'll see how useful they really can be. Their success would avoid plenty of ethical problems. In the meantime, the problem of having to destroy embryos prevails.

SNAKE OIL AND MAGIC INJECTIONS

It is important to draw a clear distinction between these promising breakthroughs and some of the stem cell therapies being peddled via the Internet. A combination of need, media hype, and unscrupulous doctors offering their services is fueling "stem cell tourism," with ill people travelling to far-off clinics on the promise of miracle treatments. Stem cells are being promoted as cures for diseases like autism, spinal cord injuries, diabetes, multiple sclerosis, or even baldness. Professionals with dubious ethical principles are willing to exploit those desperate enough to try anything. Treatments are expensive (typically in excess of $25,000 for an initial set of six injections to treat spinal cord injuries), and there is no evidence that they have any benefit.

In May 2009, the Chinese government had to tighten legislation to control the growth of unproven stem cell treatments offered at up to 150 clinics in the country.[28] It has been estimated that more than 4,000 people have already been treated there using adult and umbilical cord stem cells. From now on, clinical trials will have to be conducted and appropriate safety procedures will need to be followed, but this is unlikely to stop those who are profiting from this business. India is another popular destination for stem cell tourists despite the fact there have only been two officially approved trials of stem cell therapies, both in 2009. Other areas with lax regulations include Thailand, Russia, the Caribbean, and Latin America.

Similar cases have also occurred in Europe and America. In 2003, a clinic in Atlanta was offering stem cells to cure degenerative neuronal diseases, at $15,000 per treatment. The clinic was forcibly shut down two years later, accused of defrauding patients out of over a million dollars. Celltex Therapeutics Corporation, in Texas, became the center of attention in 2012, when the U.S. government decided to ask for proof that the stem cell treatments it was offering were safe and useful.[29] Celltex had previously received support from the state governor and health regulators, despite not having shown any scientific evidence that their treatments worked. This was the first serious intervention against these clinics in the United States, and it

THE RISK OF INJECTING STEM CELLS

Injecting stem cells or iPS directly in the body can have unexpected side effects. One of the problems is that these cells share many features with cancerous cells, and it has been shown that stem cells can replicate uncontrollably and start forming tumors. Several cases of these complications have already been described. In 2009, it was reported that a child treated with fetal stem cells had developed brain and spinal cord tumors. In 2010, a patient who received a transplant of her own adult stem cells into the kidney at an unauthorized clinic in Thailand developed an unusual mass of blood vessels that led to bleeding and forced the removal of the kidney. The patient later died from an infection when the other kidney also failed.

prompted an immediate response from Celltex, with the company threatening to move their businesses to Mexico.

In 2008, a patient accused Austrian urologist Hannes Strasser of charging him €11,000 for a stem cell treatment to cure incontinence. According to the patient, Strasser told him that the treatment had a high success rate and had been studied in clinical trials. There had indeed been trials, but the data was later found to have been falsified.[30] In the United Kingdom, Dr. Robert Trossel was investigated by the General Medical Council for offering unproven stem cell therapies for multiple sclerosis and lymphoma. In 2010, he was accused of making "false, misleading and dishonest" claims.[31] And a notorious German clinic called Xcell, was closed in 2011 following tightening of European legislation on experimental therapies.[32] For the previous four years, they had been charging €26,000 or more to inject stem cells directly into the brain or the spinal cord, claiming that they could cure or ameliorate neurological diseases but without having even attempted to prove the effectiveness. An 18-month-old baby died after being treated at the clinic, and a 10-year-old boy suffered serious complications.

Attempts to offer the public a better quality of information on these issues online have finally been established. In 2010, the International Society for Stem Cell Research launched a Web site that evaluated treatments offered around the world, highlighting false claims made by some clinics. The Web site was later closed amid threats of legal actions by several companies, which the Society feared they had no funds to fight in court.

In reality, patients signing up to these unofficial treatments are volunteering for a series of unethical and illegal experiments conducted without any of the controls that would normally be in place. In most cases the best possible outcome would be for these treatments to prove useless. Sadly, there may be serious and life-threatening complications (see box: The Risk of Injecting Stem Cells).

THE DEBATE

IN FAVOR:

- The potential to cure many diseases and not just deal with the symptoms.
- Transplants that don't require the death or inconveniencing of donors.
- No donor shortage and no more waiting lists.
- The end of immune problems. No more organ rejections.

AGAINST:

- Likely to be a costly and complex procedure, unlikely to become universally available, and hence reinforcing a rich/poor divide.
- Could require the destruction of embryos, unless a viable alternative is found (like iPS cells).
- In the wrong hands, this technique could lead to abuse: from "embryo farms," grown only to harvest cells, to the possibility of a rogue scientist succeeding in fully cloning a human (see Chapter 2).
- In keeping with other techniques that extend lifespan, might contribute to overpopulation (see Chapter 5).

CHAPTER 4

How to Improve Yourself

LIFE IN THE FAST LANE

As Charlotte warmed down from two hours of track work, her coach Justin jogged up alongside.

"Good work, honey," he said. "I gotta shoot over to the drug store. Grab yourself some lunch and I'll see you at the gym in an hour."

He then gave her a peck on the cheek. To the casual observer this might have seemed overly familiar, but no one at the track that day blinked an eye—everyone knew Charlotte and Justin had been an item for several months.

Despite an intensive workout that morning, Charlotte felt good. With the Olympics now a few weeks away, she was in the best form of her life. Barring any disaster, a medal was a certainty. If it wasn't for the fact that she'd made a decision never to read the papers in the lead-up to major events, Charlotte would have seen that most of the athletics journalists went further: for them the gold was hers. It was just a formality.

* * * * *

Six weeks later, the competitors for the final of the 100 meters are preparing themselves in the infield area just before the race start. Some are pacing up and down, but Charlotte sits contemplatively—this has been her routine in all events for several years. Today of all days she's not going to change that pattern. She has been drawn in lane 4. She's pleased: it's a good lane to have. She is confident of victory . . . or so she's trying to convince herself.

At the back of her mind a slight doubt flashes momentarily, before the psychological bodyguards get it pinned down and drag it away. The buzz around the Olympic village in the last couple of days has not been about her. The chatter has been about the unexpected performance of a Chinese athlete, Na Ling. Before the event no one had really heard much about Ling, but she'd eased through the heats and won her semifinal comfortably. To-day she will be running in lane 7.

The starter calls them to order. They wave to the crowd as they hear their names and then sort themselves into their starting blocks. On your marks . . . set . . . the gun fires.

Charlotte gets away cleanly, runs a good race, and starts the final stretch with a clear advantage. But with the finishing line in sight, she has an aware-ness of someone else to her right. Na Ling has dipped before her, she has literally edged her out at the post.

Charlotte ends up with the silver medal. She's done her best, she ought to be pleased, but instead she is gutted. She tries not to show it. Theatrical embrace with the winner, commiserative pat on the bottom from her friend Jess (who trailed in last, but is elated to have made it to her first final). Then it's off to run the gauntlet of the national and international TV crews lined up at the edge of the track.

"How did this Na Ling suddenly get so good?" she catches herself thinking.

*   *   *   *   *

After an exhausting evening of press interviews, Na Ling returns to her room, her gold medal still hanging around her neck. She closes the door, switches off her BlackBerry, and collapses onto the bed. She is experiencing a strange mix of emotions. The events of the last few days have thrust her into the global media spotlight—she knew victory would bring huge atten-tion, but the reality is far more intense than she thought. And she certainly hadn't expected a telephone call from the president congratulating her on the race!

Given her rapid appearance onto the scene, there had been a quizzical tone to some of the press coverage, with suspicion that she was taking per-formance-enhancing drugs. They can speculate as much as they like; she knows the truth. During the competition she'd given samples for drug

VOCABULARY

Bioconservatives: people who oppose transhumanism, or the philosophy that it is our duty not to accept the limitations that biology has imposed on us.

Bionic: related to any combination of biology (bio-) and electronics (-nic). Word invented by American doctor Jake E. Steele in 1958. It can refer to any object designed with the idea of imitating forms found in nature. In medicine, it specifically means the replacement of body parts by mechanic prosthesis.

Cyborg: a living organism that has been enhanced by adding technological parts to its biological base. The term was coined in the 1960s.

Doping: in sports, taking banned substances in order to bolster performance. It can be considered a form of enhancement.

Enhancement: any intervention performed to a healthy individual, not aimed at curing a disease. Opposed to treatment.

EPO: Erythropoietin, a natural hormone that stimulates the body to make more red blood cells, the ones that transport oxygen. A synthetic form of the hormone has been widely used as an illegal enhancer in competitions.

Gene doping: a technique that delivers genes to cells and can be used to help the body make a protein that enhances performance. It uses the principles of gene therapy. Due to technical difficulties, it is unlikely that it has already been used.

Gene therapy: a new medical technique that attempts to cure diseases by delivering a gene into cells that don't work properly. Despite its great potential, there are several technical problems currently limiting use of this approach.

Hyperbaric chamber: an artificial high-pressure environment that athletes sometimes use instead of physically moving to high altitude to train.

Hypoxic chamber: an artificial low-oxygen environment that athletes sometimes use instead of physically moving to high altitude to train.

Posthuman: a human that has been modified or enhanced.

Steroids: in doping, synthetic hormones derived from testosterone that are used to generate muscle mass. Nandrolone and testosterone are the most commonly used.

Transhumanism: philosophical theory that proposes that we need to use all the resources available to improve the abilities and capacities of humans. Those who oppose transhumanism are sometimes called **bioconservatives**.

Treatment: any intervention performed with the intent to cure a disease or condition.

testing on two occasions—once during the heats and then after the final. She's confident that they'll find nothing. They were barking up the wrong tree if they thought she'd been popping steroids. The assistance she had received from the doctors back home was much more sophisticated than that. Using the latest gene transfer technologies, Ling had boosted her muscle growth and the blood flow to those muscles. The drug tests will turn out clean: there have never been any artificial chemicals in her body. Everything she had gained was available naturally, provided you have the right genetics. So she wasn't really cheating . . . was she?

She'd been reflecting on that question ever since she'd had the course of **gene therapy** at a Beijing clinic. Her initial fear was that the advantage the injected genes were giving her was unfair, but the doctors on the project had reassured her that this was not the right way to think about it. She had been persuaded by their argument when they pointed out that she had only been given an extra dose of naturally occurring genes. Some lucky people already have good genes from birth. They had done nothing to earn those genes, so how was *that* fair? Her doctors had simply leveled the playing field by giving her the opportunity to be like those guys, to obtain what Nature had denied her. It was like one of those stock car races: if all the cars are the same, then the winner must genuinely be the best driver. The world could be different now. Winning could be the mark of the true sportsman, not just of the one with the best inherited DNA. They'd merely removed the element of chance.

Despite convincing Na Ling that she'd doing nothing wrong, the doctors stressed that it was important that she didn't discuss it with anyone outside of their immediate circle. They reminded her that some people in the athletics world clung to very old-fashioned views about these kinds of interventions. They would even call it **gene doping**. Ling wouldn't want to inadvertently rattle their cages, would she?

She had agreed with the doctors completely, right up until today. She won. She was the Olympic champion. Fame and fortune beckoned. So why did her victory feel so hollow?

THINK ABOUT IT . . .

Should athletes be allowed to take performance-enhancing drugs?

Is gene doping really different from taking chemical drugs? Has Na Ling simply "leveled the playing field" by altering the genetic odds in her favor?

Would you do it? Would you feel happy to win under these conditions?

Are there other aspects of life where you might be tempted to use a technological enhancement if it was available to you? How far would you go to become the best at what you do?

SIMPLY THE BEST

In recent years, the sight of a victorious team raising their trophy and spraying champagne over each other has become a standard finale to many of the major sporting contests around the world. Almost certainly their fans will be singing along as the stadium speakers blare out Queen's "We are the champions," the Black-Eyed Peas singing "Tonight's gonna be a good, good night," or Tina Turner belting out "Simply the best, better than all the rest."

To be on the receiving end of such adulation must be a great feeling. It is the ambition to be "simply the best" that has probably driven them to train for long hours, six days a week, over many years. It's also the motivation that pushes some athletes to seek a little extra help in ways that are considered to be outside the rules.

When Canadian sprinter Ben Johnson was disqualified and stripped of the 100 meter title at the 1988 Olympics, he was not the first person to try to get an edge over his rivals with the help of artificial enhancements. Athletes from previous generations have employed a weird array of concoctions in order to try to improve their performance. In one of the best known cases, which occurred in an era before regulations to curb such practices were introduced, Thomas Hicks won the 1904 Olympic marathon with the assistance of raw egg, strychnine injections, and shots of brandy administered during the course of the race.[1]

However, the fact that Johnson used a banned substance in one of the planet's most eagerly anticipated sporting challenges shone an unprecedented light into the murky underside of sport. Since that time, the public has had a much greater awareness that some athletes are using performance-enhancing drugs, a practice known as **doping**. Today's users of illegal performance aids exploit a sophisticated array of chemical and physiological interventions. Many of these have been developed for genuine clinical reasons, but can also be used to offer a competitive edge in the sporting arena. Over the past 20 years, a steady stream of competitors in a wide variety of sports has been found to be guilty of doping.[2] Cycling has been particularly marred by such accusations, to the point that controversy about who has and has not cheated in the Tour de France seems to have become part of the annual ritual of the event (see box: Cheats on Wheels).[3]

Success at sport requires a number of different factors, including peak physical and mental readiness. For most competitions, the key physiological requirements are having as much of the right sort of muscles as possible and being able to provide those muscles with adequate quantities of nutrients for the duration of the event. It is for this reason that, although some chemicals could be used to try to improve the competitor's state of mind, the majority of doping strategies are simply aimed at developing muscles or delivery of oxygen to them.

I WANT MUSCLE

Anyone old enough to have watched athletics in the late 1970s and into the 1980s will no doubt recall that track and field events were often won by rather butch-looking women from East Germany. Such was their dominance, in fact, that many of the records set in that period have yet to be bettered, despite general improvements in both equipment and training regimes during subsequent decades. Information that has come to light following the fall of the Berlin Wall has confirmed the suspicion held at the time that the success of the East German women—and their manly appearance—was due in no small measure to the widespread misuse of **steroids**.[4]

CHEATS ON WHEELS

Cycling seems to have more than its fair share of drug-related controversies. Just for starters, it has been proven that eight of the nine athletes that took the podium of the Tour de France between 1999 and 2005 were using illegal enhancements. In 2006, the Tour was marred by not one doping scandal but two. On the day before the race was due to start, the field was stripped of many of its competitors including the joint favorites—former winner Jan Ullrich and the previous year's runner-up Ivan Basso—amid accusations that there was a trail of evidence linking them to Spanish doctor Eufemiano Fuentes. Fuentes had earlier been caught red-handed with equipment and substances needed for illegal performance enhancement. The embarrassment of the sport was compounded at the end of the event when initial winner Floyd Landis was found to have abnormal levels of testosterone and was later disqualified.

Something similar happened in 2010, when Spaniard Alberto Contador was found to have used clembuterol, a stimulant that increases oxygen flow to muscles. Contador insists that the explanation is that he inadvertently ate meat that contained the drug. The story of Lance Armstrong is probably the most well known. He won seven consecutive Tours de France but in 2012, after years of investigations, it was found that he had used drugs all the time. Armstrong admitted being guilty in 2013 and was stripped of all his titles.

The steroids most commonly used in sports doping are synthetic derivatives of the male hormone testosterone (see box: The Testosterone Test).[5] Testosterone naturally fulfills two functions in men: it has a pivotal role in the development of male body characteristics and promotes the production of muscle. Because of this last property, steroids have legitimate clinical uses—for example, to aid muscle regeneration after surgery. This is an example of the tension between **therapy** and **enhancement**, an issue that we will consider in more detail shortly.[6]

Bodybuilders, weightlifters, and sprinters have a particular history of misusing steroids to bulk up beyond the natural limits of their bodies.[7] In an anonymous survey of gym users in Germany, 13.5 percent reported personal use of anabolic androgenic steroids.[8] Most of these compounds have proper medical applications, so they would be on a list of known substances

THE TESTOSTERONE TEST

Since men naturally produce testosterone, how can a male competitor be found guilty of a steroid doping violation simply by having testosterone in their body? Doping can sometimes be suspected on the basis of straightforward testosterone concentration—there are documented cases, for example, of bodybuilders injecting themselves with 40 times the normal dose of steroids. Since 1983, however, the so-called *T/E ratio* is used. It compares the amount of testosterone (T) with the amount of another compound called *epitestosterone* (E). Epitestosterone also occurs naturally in the body but is an inactive variant form of testosterone. Normally T and E are found in approximately a 1:1 ratio. If a competitor is found to have large T/E ratio it is evidence that they've received extra amounts of the active form.

and it is possible for antidoping agencies to look for them. Imagine a situation in which there was a different sort of steroid, one designed entirely for the purposes of cheating at sport. If the authorities didn't know of its existence, they wouldn't have a test for it.

This was the logic being followed by the chemists at the Bay Area Laboratory Co-Operative (BALCO), a sports nutrient company based in San Francisco. They came up with a steroid called tetrahydrogestrinone, commonly known as THG.[9] THG was a "designer" steroid, developed specifically for the sports market with the deliberate intention of evading detection by the antidoping authorities. Without a reference sample against which to compare it, and with no reason to suspect it existed, there was no way of testing for THG. Because of this it was nicknamed "the clear" by users.

It's worth noting in passing that the fact that THG was developed entirely in secret means that there is even less known about the safety of this chemical than about some of the other substances misused in sport. Despite this, during the early 2000s, several famous and successful competitors from a variety of sports became customers of BALCO. These included Marion Jones, who won five medals at the Sydney Olympics in 2000; fellow sprinters Tim Montgomery and Dwain Chambers; American footballer Bill Romanowski; and baseball legend Barry Bonds, who still holds the records for both the most home runs hit in one season and during a career.[10] It has never been proven that all of these individuals knowingly used THG, but in some cases clear evidence was eventually found. The existence of this chemical may never have come to light except that in summer 2003, the U.S. Anti-Doping Agency received a tip-off and a used syringe purported to have contained the drug. As a result of this, it was possible to develop a test for THG. It has now been added to the battery of compounds for which the antidoping authorities are screening athletes in and out of competition.

A BREATH OF FRESH AIR

It's no good having lots of strong muscles if they don't receive the amount of oxygen they need to function properly. Oxygen is delivered from the lungs to the other tissues of the body by the red blood cells. In principle, there are three ways that you could get more oxygen to your muscles: either blood flows more quickly (which is actually what happens during physical activity), you have more red blood cells, or you have more blood vessels to carry more blood (and therefore more oxygen).

Over the years, athletes—especially those participating in endurance events such as long distance running, cycling, or cross-country skiing—have found ways to increase the relative amount of red cells within the blood. Surprisingly, not all of these strategies are considered illegal; the use of altitude training has, for example, been an accepted way to achieve this aim. At higher altitudes there is less oxygen in the air than at lower altitudes. If athletes train at over 7,200 feet above sea level for a period of four weeks or more, their bodies acclimatize to cope with the "thinner" air by increasing the relative number of red cells in the blood.[11] Provided they time their return to lower altitude so that their red blood count has not normalized before they compete, the athlete takes this extra oxygen-carrying potential with them into the event. A similar effect can be achieved without physically moving to high altitude by training instead in low-oxygen or high-pressure environments in what is called a **hypoxic** or **hyperbaric chamber**, respectively. This too is considered a legitimate practice (see box: Djokovic's Pressure Cooker).[12]

Other methods for increasing the quantity of red blood cells are, however, not approved. It is illegal for an athlete to receive a blood transfusion for the purpose of boosting their performance, a practice known as *blood doping*.[13] The benefits of blood transfusion for athletic performance have been known since at least 1947 and, despite the ban, blood doping is known to have been widely employed. In particular, an athlete can have up to four units of their own blood removed at a time when they are not involved in active competition, to avoid the risk of complications with the immune system if somebody else's blood is used. The red blood cells can be extracted and stored while the other blood components are put back. The

---

DJOKOVIC'S PRESSURE COOKER

In 2011, looking for an extra edge that could keep him in the number one slot, tennis player Novak Djokovic turned to the CVAC pod, an egg-shaped pressure chamber that simulates the effects of high altitude in the body. It is claimed to be twice as effective as blood doping . . . and so far it is legal. It costs $65,000 and there are only 20 units in the world.

---

body steps up the production of these cells to cope with the loss and restores the concentration to the original level, just as it does whenever a regular blood donor makes a donation. A few days before the start of the event, the saved red cells can then be defrosted and injected back into the athlete, giving them a temporary superconcentration and with it the ability to carry more oxygen. One of the downsides of this type of blood doping is the need for the athlete to reduce their training schedule after they have stored the cells and before the body has made up for the loss.

An alternative approach involves stimulating the body to overproduce red blood cells. Central to achieving this has been a hormone called erythropoietin, **EPO** for short, a hormone normally produced by our bodies. In 1987, it became possible to buy lab-made EPO.[14] Given that EPO is a natural hormone, it was initially impossible to detect. Again, good fortune in finding athletes or their associates with materials for doping played a pivotal role in catching cheats. The discovery of substantial quantities of EPO in the possession of Willy Voet, the coach of the Festina cycling team, en route to the 1998 Tour de France led directly to the setting up of the World Anti-Doping Agency (WADA). Over time, it became possible to distinguish subtle differences between natural EPO and the lab-produced variety, allowing the cheats to be identified. This ability to spot the illegal use of hormone supplements has also led to a return to the older method of blood doping by transfusion and explains the significance placed upon Dr. Fuentes being found with transfusion equipment and packets of stored blood in the buildup to the 2006 Tour (see earlier box, Cheats on Wheels).

BETTER THINKING THROUGH CHEMISTRY

It's not only within the world of sport that people are looking for ways to bolster performance. Military commanders might be looking for ways to

WAR ON DRUGS

Chemical enhancers have already been used on soldiers. In the first Gulf War, pilots were regularly taking amphetamines to stay alert for longer periods of time and perform better in stressful situations. Some became addicted to them, and this practice was forbidden in 1993. Since then, other pills have become popular among pilots and flight physicians sometimes prescribe stimulants to fight fatigue.

In 2002, Canadian soldiers in Afghanistan were fired upon by two U.S. fighters. The pilots thought the Canadians were enemies and interpreted their maneuvers as a hostile attack. Four men died and eight were injured by the friendly fire. It was later revealed that the pilots had been taking amphetamines. It has been proposed that this could have made them especially aggressive and paranoid, two well-known side effects of these drugs, and caused them to make the critical mistake.

make their fighting men more effective in battle—making them stronger, keeping them active longer or more alert (see box: War on Drugs).[15] In education, students desperate to cram for their exams or to meet a looming deadline might take advantage of chemical support if they thought it would help them concentrate or to retain facts more efficiently. These pressures are not new, of course. Not that long ago, people routinely took concentrated caffeine tablets such as ProPlus to help them stay awake and "pull an all-nighter" if they were behind on their work.

Recent times, however, have seen greater sophistication in the choice of available drugs. The type of substance pictured in the film *Limitless* (2011), where one tablet will instantaneously unlock the full potential of your mind, does not yet exist. But there are a whole range of other chemicals—legal and illegal—that are able to alter function of the brain. Different compounds can alter our perception, "sharpen" our wits, reduce anxiety, assist in information recall, or aid concentration. For instance, methylphenidate, a drug more commonly known as Ritalin, is useful for treating attention deficit and hyperactivity disorder, particularly in children. Alternatively, if a patient were suffering from narcolepsy (excessive daytime sleeping) or had their sleep pattern scrambled as a consequence of excessive shift work or international travel, they might be given Provigil (modafinil), which promotes alertness. Some students are also taking modafinil to keep awake longer as well as Ritalin, in the belief that it will help them concentrate.[16]

The prescription of these medicines has increased significantly in recent years, beyond what might have been expected for clinical purposes. This suggests that an important number of people are getting hold of the medicines for other reasons, so-called off-label uses. Is it wrong for someone who already has normal levels of concentration to take the same drug as those with a medical problem, in order to unnaturally augment their attentiveness? There are those who think so, and this is the reason why there have been calls for universities, including Cambridge, to perform pre-exam drug tests.[17]

WHY NOT?

The distinction between therapy and enhancement is one of the key battlegrounds in debates about potential interventions to improve someone's performance. A **treatment** would be directed at curing a disease. An *enhancement* would try to improve the characteristics of a healthy body. While most believe that the first should be achieved by almost any means necessary, the same is not necessarily true for the noncurative enhancements. Why, say the enthusiasts, should one person taking a medicine be committing an offence while somebody else is actively encouraged to take the exact same tablet? Skeptics would counter that all medicines carry with them a certain degree of risk—this may be justified to cure a condition but not for nonessential purposes. It can then be argued that these drugs are available, relatively cheap, and not that harmful if taken carefully, and that the benefits could be important for all (see box: Famous Cokeheads).[18] Using Ritalin could be seen as similar to drinking coffee, a legal drug that most people take without thinking twice about it. Aside from the taste, some people overtly use coffee to improve alertness and allow their brains to perform better, especially in the morning or late in the day. It doesn't seem to bother us that it is addictive and has some important side effects. So if coffee is okay, why can't we use other enhancing drugs too?

There are compelling reasons to be careful when using any sort of enhancers. For instance, some studies on Ritalin and Provigil claim that they can cause permanent changes in brain chemistry that could be associated with mental illnesses (there is also research suggesting that taking Ritalin actually has little benefit if your concentration is already normal).[19] Going back to the sports field, a large increase in the number of red blood cells, either as a consequence of blood doping or taking EPO, can make the blood too viscous. It no longer runs smoothly around the body and you may even end up with *less* blood, and hence less oxygen, reaching the peripheries of the body. The transfusion route to increasing the red cell count

FAMOUS COKEHEADS

If it weren't for cocaine, psychoanalysis would have probably never been invented. Sigmund Freud became addicted to the drug in the late 19th century, after having used it to treat some nasal problems he had. He realized that it calmed his stress, and it is thought that it influenced some of his famous theories. Around the same time, William Halsted, an American doctor who revolutionized surgical techniques, was also heavily addicted to coke. He believed the drug helped him work more and better. In their defense, cocaine was then seen as a harmless drug. Nevertheless, both Freud and Halsted tried to hide their addictions, even when it was seriously affecting their personal lives. Would they have been as successful in expanding human knowledge without the drug?

runs additional risks of infection and the introduction of air or clots. The use of blood donated by another person can lead to infection and rejection risks. Steroids have a long list of side effects, some life threatening, but the most famous probably is a tendency for being violent (sometimes known as "roid rage"). Other chemicals come with their own problems, which sometimes include a tendency to cause addiction.

Enhancements could indeed be more harmful than some people believe. Risking health for only a marginal increase in performance is a decision that many find hard to justify, while others are happy to take the risk in exchange for a few hours of being more productive. Will there ever be enhancers that truly pose no risk to our health? In the meantime, should we keep banning them for all uses? Or should we take advantage of them with the proper controls?

IT'S IN THE GENES

The tension between therapy and enhancement remains in evidence as we move on to consider alternative strategies for improving performance. Let's begin with the role of genetic improvements, which featured prominently in our story. On the evidence as it stands, Ling would not have won the gold medal before doctors injected her with extra genes over and above her natural genome. The technique being exploited here is *gene therapy*. This technology has primarily been developed to help overcome some

GENE THERAPY GOES TO THE MOVIES . . . ALLEGEDLY

Several blockbuster films have used gene therapy to explain developments in their plots, but their scientific accuracy has been more than a little off. In the James Bond story *Die Another Day* (2002) a Cuban gene therapy clinic is alleged to be offering new identities based on the "transplant" of DNA taken from runaways who won't be missed. *Rise of the Planet of the Apes* (2011) does a little better; scientist Will Rodman is at least working on a plausible application of gene therapy, to try to cure Alzheimer's disease. However, the plausibility rapidly unravels when chimpanzees treated with the experimental medicine develop human brain characteristics overnight.

illnesses by giving the patients an extra copy of a gene that is not working properly. It's a simple enough theory and there are examples where gene therapy has worked, notably in treating children suffering from certain forms of immunodeficiency. But these have proven to be the exception rather than the rule, and so far the success rate for gene therapy has turned out to be much more disappointing than expected. This hasn't stopped people taking seriously the prospect of someone being given a genetic enhancement using this sort of treatment. WADA, who we said polices the use of illegal methods to improve sporting performance, is certainly concerned that athletes might try to gain an advantage by receiving additional genes, which they are calling *gene doping*. Gene doping is defined by WADA as "the non-therapeutic use of cells, genes, genetic elements, or of the modulation of gene expression, with the idea of enhancing athletic performance."[20]

What kinds of genes might be considered for this sort of doping? As expected, the most likely candidates are those that would enhance muscle development or improve delivery of oxygen. For muscle development, the genes most often mentioned are those responsible for making growth hormones. For increased oxygen delivery we could turn to the gene that makes EPO itself or to the gene that codes for VEGF, a protein which induces the formation of blood vessels.

Is gene doping already being used by athletes?[21] Several experiments conducted using animal models have established the principle that gene doping would work. For example, a study published in 1998 described an increase in the muscle bulk of mice when they were injected with the gene for a human growth factor. In November of 2009, researchers injected a

---

COMING SOON?

The first product that might be exploited for gene doping in athletes is already in advanced stages. Repoxygen is a gene therapy system being developed by British company Oxford Biomedica to treat anemia. It is basically a mechanism to deliver into cells the gene that makes EPO. Injected directly into the legs of lab mice, it has been shown to increase the levels of EPO in the muscle. There have not been any tests in humans yet, so it is unknown whether it would work and how safe it would be. Nevertheless, there are fears that some athletes may have already been tempted by Repoxygen.

---

gene for follistatin into the legs of monkeys and saw an increase in the size and strength of the monkeys' muscles after just a few days.[22] They hoped to be able to use this knowledge to eventually treat certain muscle-wasting diseases. Because the clinical success in trials of gene therapy has been so limited, it is believed that no athlete has yet been the recipient of gene doping. There is, however, a feeling that major breakthroughs in the legitimate uses of gene therapy are very close, in which case applications aimed at enhancement are likely to follow (see box: Coming Soon?).[23]

MARCH OF THE CYBORGS

After chemistry and genetics, the next level of enhancement could be adding mechanical pieces to our bodies. Hybrids between humans and machines, sometimes called **cyborgs** (from *cyb*ernetic *org*anisms), are again a familiar feature in science fiction stories but remain a distant reality. Connecting the human brain so as to control any robotic extensions has been seen as a major stumbling block, but here too genuine progress seems to be occurring.

In the summer of 2011, President Barack Obama awarded Sergeant Leroy Petry the Medal of Honor for bravery on the battlefield. In a striking photo taken at the ceremony, the president is shown shaking hands with Sergeant Petry's grey **bionic** hand, which replaced the one he had lost saving his comrades from a grenade.[24] In 2002 British Royal Marine Jim Bonney had his leg amputated following a climbing accident. He has been described as the first serviceman to return to active duty after receiving a prosthetic limb.[25] On some levels this claim does a disservice to countless

INTELLINGENT LEGS AND HANDS

Prosthetic legs are more advanced every day. At the end of the last century, the mechanical ones became popular. They can bend, but users have to walk using an awkward gait to avoid falls. The solution is the intelligent prosthesis, like Otto Bock's C-leg. It contains a microprocessor that controls the way the artificial knee bends, making it behave almost like the real thing. It still requires quite an investment of energy to walk on one of these. Bock also designs the Michelangelo hand, an articulated electronic prosthesis that is controlled by the arm muscles and has a very realistic look. Another alternative is the iLimb, made by Touch Bionics, which was the first in this category. Unfortunately, the cost of any of these prostheses is around $40,000 each.

sailors of the 18th and 19th centuries who continued to fight despite having wooden peglegs. In sharp contrast to those earlier amputees, Captain Bonney has a set of exchangeable prosthetic limbs that he can select according to task. He can choose one for running, one for water-based operations, one for long marches, etc. Although he still struggles to be able to perform as well as a soldier with two intact legs, it is possible to imagine a situation where these artificial extremities could eventually perform better than real ones. Some could even have weapons attached. Will we see cyborg soldiers in the battlefields one day?

The use of mechanical aids is also becoming an issue in the sporting arena. In particular, the case of South African runner Oscar Pistorius brought sharp debate about the use of prosthetic limbs. As a consequence of a congenital defect, Pistorius had both legs amputated below the knees when he was a child. For several years he competed successfully in the Paralympics using J-shaped carbon-fiber legs, as a result of which he was nicknamed Blade Runner, and was sometimes described as "the fastest man on no legs." At the height of his success, Pistorius started to run so well that he achieved times that allowed him to qualify for mainstream athletic events (including the 2012 Olympics, the first time an amputee would compete in such event). Although as an individual he was not achieving the same times as the top runners, he actually won a silver medal at the 2011 World Championships as part of the South African 4x400-meter relay team (he ran the heats but was dropped from the actual final). At the 2012 Olympics, in London, he was last in the 400-meter semifinal and his team came in eighth in the relay final. Unfortunately, Pistorius achieved a different kind of fame when he shot his girlfriend dead on Valentine's Day 2013. He insisted it was an accident, but was eventually charged with premeditated murder. It remains to be seen if he will be able to return to athletics after his release.

The involvement of Pistorius alongside able-bodied athletes remains controversial. Questions have been asked about whether the design of his prosthetic legs actually gave him an unfair advantage. In fact, he was banned from competing in 2007–2008 while the properties of his legs were assessed. Several aspects of their performance were then investigated. First, the legs are lighter than normal legs. Second, they have a certain springiness achieved from the flex of the leg. Third, because he does not have to provide oxygen to as many muscles as a regular athlete, he is not as likely to suffer from the buildup of lactic acid (which gives the infamous muscle pain after exercise). Fourth (and this has not been leveled at Pistorius himself), there is the potential for someone using artificial legs to be fitted with limbs that make them taller, and with a longer stride, than they would naturally have had. In other research, it has been shown that Pistorius's stride is in fact shorter than that of regular athletes (he needs 250 steps per minute instead of an average 240 for able-bodied sprinters). His starts were slower than theirs and the spring effect is not as markedly different from the function of regular ankles as had been imagined. In addition, running with blades is difficult and painful. Opinion of the experts, therefore, remains divided.

Researchers in the field of artificial limbs are now looking into the next generation of products. The goal would be to connect the prostheses to the body through some sort of interface that links them to the nerves. They use fiber optics and nanotechnology to bypass the problem of metal parts rusting and degrading once they are inside the "moist" environment of the human body. This may allow the amputees to both move the hand or leg and also use it to "feel." We are clearly far away from this situation, but will prosthetic limbs improve enough to really become an advantage one day? Could amputees ever outrun the fastest athletes? Could having normal legs eventually be considered a *disability* in certain professions (athletes, soldiers . . .)? Could somebody willingly elect to replace one of their extremities by a mechanical one (see box: The Hand of God)?[26]

WHO WILL BE THE FIRST BIONIC MAN?

Without having to go that far, bionic extensions could also be *added* to our bodies. Electronic chips have already been implanted under the skin in different tests. They can record and store data and could be useful from a

THE HAND OF GOD

Patrick Mayrhofer lost his left thumb and the use of his right hand when he touched a live wire in 2008. After a complex surgical process, which included transplanting a finger from his other hand, he was able move it again. The problem was that then his left hand only had three fingers remaining. So he decided to have it amputated and get an electronic hand instead. This raised a series of moral concerns that prompted a meeting of 80 experts (from the fields of medicine, law, ethics, and theology) to decide if he should be allowed to do it. Some saw it as replacing a hand "made by God" with one made by humans. Eventually, he was authorized to go ahead.

medical point of view or even to identify and track a person. Controversial U.K. engineer Kevin Warwick[27] has at various times implanted a radio transmitter into his arm and, more spectacularly, set up an electronic network directly from nerves in his forearm that allowed him, by the power of thought alone, to control a prosthetic arm on the other side of the world. American body artist Steve Haworth once inserted a magnet in a customer's finger, which allowed him to "sense" magnetic fields (the magnet would vibrate in their presence).[28]

Could small implants like these be the first step to true bionic enhancements? Could they be a way to control our environment and even expand our traditional senses? Permanent insertion of metallic devices in the body is still not completely safe, since the risk of infection and other complications is very high. Warwick's first implant, for instance, was removed after only nine days. The second one was even riskier, since it could have affected his nerves and cause chronic pain or even loss of sensitivity and the ability to move his hand. He had it for three months and, at the end of this period, the implant was already beginning to affect his functions. The problem of a two-way "rejection" between the body and the implant will have to be solved before we think about mixing flesh and metal permanently.

One solution could be a new type of "electronic tattoo," developed by scientists in 2011.[29] These are circuits printed in ultrathin flexible materials, with properties similar to those of human skin. They call it "epidermal electronics," and they can be grafted onto skin just by rubbing them with water, like they were temporary tattoos. For now, they can remain attached for only 24 hours, but this performance could be increased in the future. So far they have been tested to collect data. When placed over the heart, for instance, they were able to record the same information that an electrocardiogram would measure. Another implant, which contained a microphone, could detect voice commands when placed on the throat. This allowed the subject to give instructions to a computer, something that could be very useful for people with disabilities.

THE END OF HUMANS AS WE KNOW THEM?

If we ever manage to solve the technical issues, enhancing our bodies with synthetic parts could prove to be very advantageous. If we take this to the limit, at some point our body could be mostly artificial. Would we then still be human? What separates man from machine? Is it self-awareness and a brain? Can we consider ourselves human if our brain is transplanted into a robot, such as in the *Robocop* movies? Is the brain the minimum required for defining humanity? What would happen then if we start adding chips to help our neurons (see box: Time for an Upgrade)?[30] We could boost our memory, our intelligence, or even the way we feel. How would that make us different from a robot built from scratch?

While we may still be years away from having to worry about setting the ground rules for such radical modifications, the debate of whether this is

TIME FOR AN UPGRADE

Computer chips can store information in a way similar to how our brain does. The only thing is that the "substrate" is different: circuits in the first place, neurons in the second. Is there a way to blend both systems? Could we "expand" our memory by adding a chip to our brain the same way we upgrade computers? Could we even implant chips with pre-stored information, like a whole encyclopedia? We would first have to solve how the biological parts could be able to access and control the mechanical ones. It's the same problem we face with other prostheses . . . and it is a major one. Whether or not this would ever be possible, it still raises an interesting question. What if one day we could transfer all the information contained in our brains to memory chips? Would they be able to think? Would they be "us"? Can a person live in a fully mechanical brain and body? Or would that be just a very advanced computer, a machine after all?

something that we should strive for or even allow is very much alive. Central to the discussion are the **posthumans**: those who will have their bodies improved by any means to a point of almost becoming a separate species.[31] Will enhancements eventually represent the end of the human race as we now know it? Is this desirable?

Some believe it is our duty not to accept the limitations that biology has imposed us. After all, one of the strengths of life on this planet is its ability to constantly evolve. This vision has been called **transhumanism**.[32] People who oppose this are often called **bioconservatives**. The obvious reason for not supporting enhancements is the side effects that most of the procedures have. Apart from this, one of the greatest dangers is to worsen the worldwide divide based on wealth, as is true for other advancements discussed in this book. Rich people would have access to all sorts of modifications, while others would not be able to afford them. This would increase the differences in survival, quality of life, and life expectancy between rich countries and the developing world, effectively creating two separate populations of humans and posthumans.

Will transhumanists or bioconservatives eventually win the battle? This is a debate that will heat up over the next decades, as scientific advances that will make extreme human enhancement become more feasible.

THAT'S NOT FAIR!

Despite official resistance to all forms of enhancements, there are a growing number of prominent voices arguing in favor of the legalization of various techniques, at least in some specific situations.[33] Over and above issues of personal choice ("It's my body, I can do what I want with it"), some opinions are based on the premise that the principle of "fair play" is not well served by the present situation, since some people are born with genetic advantages that make them better at certain things. For instance, many sports are self-evidently easier for people of a particular physique. This was the argument Na Ling's doctors had put to her to justify gene doping in our initial story. There is often significant advantage in being tall. Basketball, tennis, rugby, and rowing are just some of the examples where height is a contributory factor in success. If you are 5′7″ tall then it is very unlikely that you are going to make it as a professional basketball player. You've done nothing wrong, you might even do more training than everyone else, but you've simply been dealt a losing hand in the genetic lottery. It's not fair that somebody who put in minimal effort but happens to be 7′2″ gets picked for the team.

WHERE'S THE LIMIT?

What sort of "improvements" can be considered okay in sports? Archers are allowed to wear glasses or contact lenses to correct deficiencies in their vision. These interventions might be seen as therapies rather than enhancements. How about laser eye surgery? Are the ethics different if these approaches can give them *better* than 20/20 vision? Is the difference between treatment and enhancement what sets the limits? How about special swimsuits or shoes, designed to improve speed and reduce friction? If these are not available to all competitors are they unfair? Or are they okay just because they do not pose any health risk for the athlete, when compared to doping?

There are more subtle examples. Eero Maentyranta, the Finnish skier who won three Olympic gold medals in 1964, was later found to have a genetic mutation that meant that he had 40 to 50 percent more red blood cells than average—exactly the sort of effect others seek to achieve by taking EPO.[34] There is also growing evidence for a genetic contribution to ability as reflected in the geographical heritage of sportsmen who excel in different disciplines. For example, runners originating from West Africa dominate in the sprints whilst athletes from East Africa are predominant at long distances. Is trying to imitate these "naturally enhanced" individuals really cheating (see box: Inhibiting the Inhibitor)?[35] Why should we forbid the use of drugs or genes to "level the playing field"?

As we said earlier, a compelling reason could be that most enhancement techniques come with an implied risk. For instance, the basic procedures of gene therapy that would provide the necessary groundwork for gene doping have proven more problematic than anticipated, including the death of participants in regulated clinical trials.[36] There is growing interest in a different method of genome editing called clustered regularly interspaced short palindromic repeats (CRISPR), based on a naturally occurring bacterial system;[37] however, it is too early to know whether this will be feasible for safe modification of humans.

Apart from the immediate risks, there could also be unknown long-term side effects, like increasing the risk of developing a cancer by using hormones to stimulate muscle growth. To sanction any sort of enhancements regardless of the possible complications would automatically discriminate

INHIBITING THE INHIBITOR

Myostatin is a natural protein that actually stops muscle generation. In 2004, the curious case of a German boy was reported in a medical article. From an early age the boy was developing muscle far faster than would be expected. Genetic analysis showed that he had a mutation in the gene that makes myostatin. As a consequence he wasn't making enough of the protein that should have been regulating muscle production, allowing him to become, in effect, a super strong child. Interestingly, his mother, who had been a professional sportswoman, also turns out to have the same mutation. Should this family be banned from competitions due to this unfair genetic advantage? Is this any different from having been born with longer legs or more endurance? And is it any different from actually blocking the gene using scientific methods if you are not lucky enough to have this mutation?

against those who cannot, or do not want to, use them. It also seems inevitable that if certain chemical or genetic methods do become legal there are still going to be those willing to pursue other illegal strategies carried out in secret, in the hope that it will give them the necessary edge over their rivals. Legalizing certain enhancements could simply shift the position, in the same way that uncovering a new doping system forces cheats to look for new ones.

THE DEBATE

IN FAVOR:

- Enhancement of the human body leads to better performances. In sports, it would help achieve better marks and make competitions more spectacular.
- There are already legal enhancers being used worldwide, like coffee. Why limit ourselves?
- We have the right to decide whether we are prepared to take the potential health risks in order to achieve our ambitions. In the case of athletes, professional training, even without enhancers, also involves risks and can have long-term effects on health and quality of life.

- Approving doping techniques could reduce the unfair advantages that some athletes have due to their natural genes.
- There are some legal "tricks" to obtain the same effects than you can get with certain drugs (like training at high altitude to boost red blood cell numbers). If the end point is similar, why should we forbid only the chemical way of achieving it?

AGAINST:

- We can't guarantee that enhancements will be safe.
- Using illegal methods to improve your performance is cheating and can be considered immoral.
- Since not everyone could or would want to use enhancers, if they were approved they would in fact make things even more unfair.
- Danger of creating a society of rich posthumans separated from the "traditional" humans.
- Legalizing certain enhancing techniques would just lead people to look for an edge elsewhere and bring novel illegal (and probably more dangerous) ways of improving themselves.

CHAPTER 5

Who Wants to Live Forever?

LIVE LONG AND PROSPER

Everything is exactly as the Web site had promised. A lavish resort, in the middle of the desert, dripping with luxuries. There's a spa, swimming pools, solarium, gym . . . everything you could wish for, just as you'd expect in these fancy places. But here they're offering more than your usual health resort. They're offering immortality.

Most of her life, Amy had been skeptical about all these antiaging businesses. She was sure there was nothing on this planet that could help you live longer. Since she'd turned 40, however, Amy found herself increasingly experimenting with the different solutions on offer, from creams to vitamins, from antioxidants to hormones. But despite the claims on the boxes and in the flashy advertisements, the wrinkles had kept coming and the only effect seemed to be a steady draining of her bank account. A few months shy of her 50th birthday, Amy was just about ready to give up and admit that you can't beat Father Time. It was then that she found their Web site.

This was completely different. It wasn't about taking pills or drinking gallons of green tea. Quite the opposite, in fact. And unlike the other treatments she'd tried before, this one really appeared to be scientifically proven. The Web site had links to very serious research describing success in extending the lifespan of all sort of animals, from worms to mice. Some of the scientific papers even showed that the treatment could help prevent cancer. Although she couldn't understand half of what they said, the studies had been published in journals that had the highest reputation. There were quotes and testimonies from lots of people with the title "Professor" in front of their name—this had to be for real. If it worked in all animals they tested, why should humans be any different?

That's why she packed her bags and headed to this revolutionary clinic set up in California by a group of visionaries right after the results from the first experiments became public. It was expensive, for sure, but the chance to extend her life was worth every penny. A week-long intensive program would apparently give her all the tools needed to achieve something very close to immortality. After that, no more money to spend, no hidden fees. She was willing to risk it. What was there to lose?

For the first few hours in the clinic, she thinks she's made another mistake. She waits in the lobby with a small group of people, mostly her age, all looking eager to get started. After a while, a very thin girl in a nice bright suit welcomes the new arrivals and shows them the facilities. Finally, the girl takes Amy to the room where she'll be staying. It looks just like any of the resorts she has been before. So far, so good.

Having unpacked, Amy spends half of the morning attending seminars led by some senior members of staff. Most of them look like new age gurus, extremely thin, with big smiles and long beards. There are a lot of mystical elements in their talks; too much for her taste. She's not into that. Can't they skip the theory and go straight to the point? She should have done some more research before spending the money, she thinks. It looks like she's stuck here for a whole week with a bunch of loonies trying to convince her to join their cult!

And then lunchtime arrives. "Lunch" though is an overstatement for what's on offer: half a plate of spinach and a glass of water. She thinks about complaining but then she remembers that this is the whole point—controlled nutrition. The promise of immortality comes with a high price tag: severe reduction in the amount of food you eat—very severe! And for the rest of your life. "Caloric restriction," they call it. That's exactly what kept those worms and mice going for much longer than usual and made the scientists get so excited. Amy doesn't know if she will be able to stick to this crazy diet for long, but she sure won't give up before trying.

In the afternoon, her belly starts to complain. The rumbles are so loud that she fears everybody can hear them. But the other members of the group are having the same problems. They don't look happy at all. Nobody wants to be the first to give up, so they all sit in silence through another round of supposedly enlightening presentations.

VOCABULARY

Caloric restriction: a severe and prolonged reduction, of at least 40–60 percent, in the amount of food we eat. It has been shown to prolong the lifespan of certain animals in the lab.

Oxidative theory of aging: A widely accepted theory that states that we age as a consequence of damage to our DNA caused by exposure to oxidants.

Reactive oxygen species (oxidants): toxic chemicals that can be found naturally in the environment, as well as being the by-product of cellular respiration.

Resveratrol: a chemical found in grapes and other fruits that has been linked in the lab to better health and longer lifespan. The mechanism is still controversial, and it is not clear what sort of dose could have an effect on humans.

Telomeres: Structures at the end of the chromosomes. They get shorter with each cellular division, thus acting as "molecular clocks" that record the age of the cell. When they shorten below a certain point, they also trigger the cellular aging program.

Dinner isn't any better: a fruit salad and half a glass of a gooey milkshake that tastes horrible. By the time she goes to bed, Amy feels exhausted, frustrated. And hungry. Very, very hungry. She's always been good at diets because she's stubborn. She can stick to a plan if she sets her mind to it. This is one of the reasons she thought this could be easy for her. But here there's one big difference: this will not get any better; this is *forever*.

In the morning, she's not feeling very good. Hunger has turned into a sharp stomachache that a glass of orange juice and an oat cookie are not able to quench. Unfortunately, they are not giving them more food until lunchtime. She can hardly think of anything else during the yoga class and the lectures that follow.

She knows it's all in her head: she can survive a few days without eating much; she has done it many times before. Every year after Christmas, for instance, or whenever she had to jump into a bikini. What kept her going then was knowing that after all the effort there would be a prize. The light at the end of the tunnel. The prize is even bigger now, it's true. Immortality, or at least a much longer lifespan. That's the idea: years and years of happy and healthy living. That should be a big enough motivation!

Okay, maybe longer and healthy living . . . but happy? Is this happiness? Is this how she wants to spend the rest of her (extended) life? Dreaming about a juicy steak? She avoids asking herself these questions for the following days. She's determined to spend the whole week there. That's what she paid for and that is what she will get.

After the third day, things get easier. She is now stuck into a routine and her cravings are not as intense. The weakness she felt is not that bad anymore; maybe just because she got used to it. The truth is that everybody looks miserable, except the organizers, who keep smiling all the time. But she can't find any reason to smile.

Finally, the day to go home arrives. They give her a book with all the instructions to continue the program on her own and wish her a long and prosperous life. On the plane, her head spins. At least she can say one thing: she lost those few extra pounds she had been carrying since last year. What's so bad about eating less? Doctors always say it's good for you. The problem is that this is closer to starvation than she thought. But on the other hand, the reward is immense . . . if it's true. Nobody knows what this sort of diet will do to humans. We are not mice, after all. If in the end this turns out to be completely useless? But what if it's not? Can she afford to ignore it while a group of hundred-year-old hippies get to come to her funeral and laugh at her grave?

When she arrives home she is really exhausted. She throws the suitcase in a corner of the room and heads to the kitchen. She takes the jar of peanut butter from the cupboard and prepares herself the biggest sandwich she's ever seen.

THINK ABOUT IT . . .

Should we try to extend our lifespans? For how long? Indefinitely? Is that fair when so many people elsewhere are dying before they reach what we would consider "middle" age? Aren't there more pressing health problems we should be dealing with?

If we eventually find a procedure to prolong life, who would have access to it? Only the richest?

Is longer life worthwhile if the means to obtain it radically reduces our very enjoyment of life? How much are you willing to sacrifice to live more?

What would be the impact on Earth if we all start living much longer than we do now?

THE FOUNTAIN OF YOUTH

Finding immortality has been an ambition of mankind for almost as long as we have walked the earth. Numerous ancient legends tell of fruitless searches for the fountains of youth. The myth lives on, given a fresh, scientific-sounding spin for the modern era. Pills and potions sold in stores and over the Internet that promise longevity are the cornerstones of an immensely lucrative market. Millions of pounds a year are invested in these treatments, despite the fact that we have no scientific evidence that they have any beneficial effect on humans. These "magic" pills are sold not only by companies of dubious reputation but also by well-known brands, like those that clutter store windows with products that have never been proved to retard aging (see box: The Ultimate Antiwrinkle Cream).[1] The fact that current laws in most countries allow these compounds to be sold as "supplements" or "cosmetics" provided they are not labeled as "medicines" (which would force them to undergo more severe controls), only underscores the political clout that this very profitable industry has throughout the world.[2]

Notwithstanding the fact that no verified treatment for ameliorating the symptoms of aging has been identified, our knowledge of *why* we age has actually increased exponentially in recent decades. The combination of new genetic techniques and advances in biochemistry are bringing us closer to understanding how our DNA gets progressively damaged and how this plays a pivotal role in the molecular basis of aging. It is also true that we have learned how to modify aging in the lab. Through chemical treatments and genetic manipulation we can make flies, worms, mice, and other animals that age much faster than usual or that survive for twice the normal amount of time.[3] This is a proof of principle—time can be beaten. Once you know why and how something happens, you are much closer to having the power to change it. And, at least in animals, we have already started doing so.

Our understanding of how aging works in humans is still far from complete. Nevertheless, we've reached a point where some experts no longer believe it's impossible to defeat aging. For them, it's just a matter of time—time until we can do it, and the amount of extra time we can achieve when we do. Many think that only active antiaging strategies will allow humans to keep extending life expectancy, which is otherwise likely to plateau around an average of 85 years.

Some scientists even take this one step further. They say that if we can completely and continuously suppress the deterioration of our cells, in theory this could lead to true immortality. Whether this would ever be a

THE ULTIMATE ANTIWRINKLE CREAM

Most wrinkle creams have little effect beyond being moisturizing, despite their sometimes astronomic prices. Some rely on principles that could be considered quite close to pseudoscience. Experts complain that some creams that contain collagen are sold as being able to restore the amounts of this molecule that we have under the skin (which would prevent the formation of wrinkles). In fact, collagen molecules are far too big to penetrate the skin on their own.

All this seems to be changing. Most big cosmetic companies are starting to prepare new products that have actually passed the proper tests. L'Oreal has LiftActive, a cream that contains a sugar compound extracted from a tropical plant that has been scientifically proven to stimulate the generation of collagen. The methods used by the researchers at L'Oreal to identify this new compound were similar to those used to find new drugs against cancer or other diseases. Boots No7 Protect and Perfect has also passed a test that seems to show it has genuine antiwrinkle properties, the first commercial cream to do so.

reality is hotly debated by specialists. In the meantime, less dramatic advances in this field could have significant impacts on medicine in a very near future.

A SIDE EFFECT OF LIVING?

So, why *do* we age? The easy answer is that aging is a consequence of being alive. We can't have one without the other, or at least this is what we thought until recently. There have been different theories proposed to explain the molecular basis of aging, including: the accumulation of toxic residues inside and outside the cells; the shortening of **telomeres** (see box: Does Size Matter?);[4] the accumulation of damage in the mitochondria (the power plants in our cells) that limit the amount of energy they can produce; or the progressive loss of the adult stem cells that are in charge of regenerating our tissues (see box: Stem Cells: The Alternative).[5] Recently, it has also been proposed that the side effects of certain processes necessary for our survival can accelerate aging. For example, some of the mechanisms that protect cells against cancer could at the same time be making our cells

DOES SIZE MATTER?

In 2011, two companies started offering the possibility of measuring the length of the telomeres within the cells of anyone interested in knowing (and willing to pay). Both companies have academic clout: Telomere Health, in California, was founded by Elizabeth Blackburn (who got a Nobel Prize in 2009 precisely for having discovered telomeres) and Life Length, in Madrid, founded by María Blasco (one of the leading world experts in telomere research). These tests work on the premise that there is some statistically significant correlation between long telomeres and staying healthy for longer. The suggestion is that telomere length serves as a marker of the real biological age of a person. Carol Greider, who shared the 2009 Nobel Prize with Blackburn, is one of many experts that believe that such tests are premature and of little value. There is still controversy regarding the relationship between general telomere length and human health, and it's not clear how knowing the exact length of your telomeres offers any life-altering benefits. In fact, scientists are still arguing over the most appropriate way to measure telomere length in the first place.

grow old faster. Apart from all these processes, we would also have to take into account our genetic predisposition to age and the effects of the exposure to different levels of the harmful agents found in the environment during our lifetime.

In fact, the same oxygen that we need to sustain life actually causes damage to our cells, resulting in progressive deterioration. This happens because the breakdown products of oxygen, known as **reactive oxygen species** (or **oxidants**), produce small but measurable damage to our DNA. The steady accumulation of these lesions, combined with the impact of the sun's rays and environmental chemicals, eventually causes our cells to age. This **oxidative theory of aging**,[6] as it's usually called, is widely considered by experts to be one of the main mechanisms responsible for the deleterious effects of time on our bodies. It's also probably the most well-known by the public due to its frequent appearance in news stories and advertisements for antiwrinkle creams! These give the false impression that scientists have already been able to pinpoint the single origin of wrinkles and grey hair.

In truth, as we have already noted, the aging process is more complex than just a direct effect of oxidation. A paper published in June 2010

STEM CELLS: THE ALTERNATIVE

Aging is, in fact, an accumulation of tissues that have stopped working the way they are supposed to. That's why our skin loses its elasticity or why our hair loses its original color. If we could find a way of replacing everything that is not working properly, we could not only cure a considerable number of diseases but maybe even find an alternative route to immortality. In an earlier chapter we discussed the possibility that regenerative medicine could produce the organs we need. This could also extend lifespan, not by addressing the causes, like antiaging procedures would, but by alleviating the symptoms. In a similar fashion, transplanting stem cells to replenish the pools of adult stem cells we lose with age is being considered by some as a possible way of fending off the effects of time on our organs. It's possible that in the near future we will see the first scientifically proven therapies that follow this principle.

showed that you can expand the lifespan of a worm (a model organism used frequently in the study of aging) without having to reduce the levels of oxidation happening in its cells.[7] The fact that antioxidants, chemicals that block the oxidants and the damage they inflict, have only limited benefits when we scale up from lab experiments on cells to studies on whole organisms also shows that there are indeed more issues at stake. Overloading our cells with antioxidants can even be counterproductive, because we actually need *some* oxidants to function properly and the body has mechanisms for redressing the balance of oxidants if we take the level too low. Other studies have shown that naturally occurring oxidants prevent cells from detaching and travelling to other places in the body (a property that is crucial in the spread of cancer). Accordingly, antioxidants may actually *increase* the chances of developing a malignant tumor.[8]

The simplistic notion that "the more antioxidants the better" is also challenged by a recent study conducted by scientists from Kansas State University that showed that antioxidants can impair muscle activity.[9] The explanation for this observation goes like this: Certain oxidants are vasodilators (i.e., they increase the flow of blood to a tissue), and since antioxidants counter the activity of oxidants, an excess of antioxidants can actually reduce the supply of oxygen to muscles to such an extent that their function will be impaired. Since the heart is, in effect, a specialized muscle, a large dose of antioxidants

---

IF WE COULD TURN BACK TIME . . .

Most antiaging strategies attempt to slow or even stop the deleterious effects of time on our cells. What if we could go even further and actually *erase* them too? Scientists working at Albert Einstein College in New York, led by Ana Maria Cuervo, are leading the way in examining how clearing of damaged proteins from old cells can prevent the functional decline associated with aging. Some of their work suggests you could not only stop but reverse cellular aging. This opens new possibilities in the treatment of aging, since there would be no need to start them at a young age: old people could be made to look 20 again.

---

might even increase the risk of a heart attack. It goes without saying that any positive effect on aging achieved via antioxidants would be of little merit if you died in the meantime from a heart disease or from cancer. The answer to longevity has to lie somewhere else (see box: If We Could Turn Back Time . . .).[10]

SHOULD WE START STARVING OURSELVES?

A few years ago it was discovered that a severe reduction in the amount of food eaten by an animal seems to affect its lifespan. This approach, which we refer to as **caloric restriction**, sparked a revolution in the study of aging.[11] It is currently the most effective way to slow down aging in an experimental context. We are not talking about cutting out the odd chocolate bar: caloric restriction involves cutting our food intake by as much as 60 percent. This seems to work in almost all the models that have been tested, ranging from simple worms through to mice. Research conducted at the Wisconsin National Primate Research Center has suggested that caloric restriction even had an effect on monkeys, our closest relatives tested to date.[12] In their experiments, monkeys fed a reduced diet for long periods survived longer than other monkeys given a normal diet (although we must point out that other researchers have not found the same response in their own studies).[13]

Would caloric restriction work in humans? We don't know for sure, but there are certainly people testing the idea, in ways similar to the protagonists in our scenario. There hasn't been a scientifically designed clinical trial yet, so it is unlikely that we will be able to draw any firm conclusions from these real-life experiments. Even if we wait a few decades, the results of un-regulated use are likely to remain inconclusive. It will be a difficult question to answer categorically, since the logistics of setting up a controlled experiment lasting for that long would be phenomenal. It would also be ethically problematic to treat the study subjects with such a severe and prolonged diet, no matter if they were initially willing volunteers. That is why at this stage, scientists are still working to define the effects of caloric restriction at the cellular level before planning on testing its effects on humans.

Although the mechanism by which the process may work has yet to be fully understood, it makes intuitive sense that reducing food intake prolongs life. If less food is eaten, then there will be a slowing of our metabolism, and consequently a reduction in the amount of harmful by-products being generated. From an evolutionary point of view, it is also a good idea: when food is scarce, you would want to prolong the lifespan of the animals in a herd as much as possible to allow them to survive until the problem is resolved.

So, eating less—provided we don't literally starve—is a good thing, right? Well, not necessarily (see box: Quality vs. Quantity?). What people fail to mention when caloric restriction is discussed is that in most of the animals tested, life extension comes also with a reduction in fecundity, i.e. our ability to reproduce. Again, this makes perfect sense: in an environment with little nutrients it is foolish to invest energy in making new animals that would have to compete with their parents for the same meager resources; it's more effective to focus on keeping alive the ones that already exist, who can reproduce later, if they survive. When it comes to marketing the benefits of caloric restriction for humans, it's easy to see how a drop in fertility might be seen as only an inconvenient side effect.

The most important thing is that researchers have discovered a series of molecular pathways that get specifically activated when cells don't receive the normal amounts of nutrients for a prolonged period of time. They are currently trying to discern which of these pathways are actually playing a role in prolonging the life of the animal. And this is the key. Nobody wants to survive for decades on a diet of lettuce leaves and water. The trick would be to activate those same pathways without having to eat less. In other words, to fool our bodies into thinking we are starving.

Some experts think that a chemical called **resveratrol**, which is found in grapes and other fruits, could be the one that will eventually do the trick. Based on very promising lab experiments that started in the first years of this century, a series of studies are currently being conducted on this drug.[14] One of the problems is that its role in aging seems more complex than was initially anticipated, and its mechanism of action is still being heatedly

QUALITY VS. QUANTITY?

Extending the use-by date for our bodies doesn't make us invulnerable to disease. Hand in hand with research to combat aging, we need to have research looking at ways to improve the quality of life we'd experience in our prolonged existence. It wouldn't be very useful to remove the barriers that prevent us from reaching our 100th birthday if we can't also find a way to ward off diseases associated with old age. The longer we live, the greater are our chances of dying via one of the main causes of death in the developed world, like heart disease or cancer. Some even believe that the current obesity epidemic in some developed countries will eventually decrease overall life expectancy if we don't find a solution first. If these diseases don't get us, we may fall foul of one of the dementias, such as Alzheimer's, that are increasingly associated with aging. There's no point living forever if you'll live out those extra years in a care-home, unable to remember your friends and family.

discussed. For instance, not all agree that resveratrol can actually increase the lifespan of *healthy* lab animals, although it seems to have a striking effect on mice being fed a high fat diet.[15] Nevertheless, resveratrol is already being sold in stores, even before we know if it works in humans and, if so, what dose would be effective.

Even better results have been recently seen with rapamycin, a compound that seriously prolonged the lifespan of laboratory mice, even when treatment wasn't started until an age equivalent to 60 human years.[16] Rapamycin is named after Rapa Nui (Easter Island) where it was first discovered. It is produced by a bacterium called *Streptomyces hygroscopicus*. Of course, the bacterium didn't make this compound to keep laboratory animals going forever, it is a fortuitous accident that it seems to have this effect (in the same way that compounds produced by similar bacterial species were not "intended" to be useful to humans as antibiotics).

Once again, however, use of rapamycin (also known as Sirolimus) is not without its complications. The drug also shuts down the human immune system, which is why we are currently more likely to encounter this drug as an immunosuppressant during transplant surgery than as an antiaging pill.[17]

It seems unlikely that chemicals such as resveratrol or rapamycin are themselves going to be the sought-after elixir of life. However, they may prove to be the natural prototype that chemists can tweak in their laboratories in order to produce a pill that retains the desired features without the side effects. As with so many of the scientific breakthroughs we are considering in this book, you can't help but feel we are on the brink of something really big, but we haven't quite reached the point where our accumulating scientific knowledge translates into real-life treatments.

ROOM FOR EVERYBODY?

Let's assume that humans do, after all, find a way to prolong our lifespan way beyond the current limits. Many important questions would immediately follow. Who would have access to this technology? Will it be available to everyone or limited by either geography (where we live in the world) and/or ability to pay? Already we know that access to essential medicines has always been unevenly skewed towards inhabitants of developed nations. This was one of the main reasons for a big divergence in life expectancy (that in 1950 was 65 years in developed countries and only 41 in poorer regions). It has been predicted that this gap will slowly close, which should lead to a difference of only 88 vs. 81 years at the end of this century. Would an ability to halt aging exacerbate again the imbalance between countries?

Should we stop developing a drug simply because it will not be equally available to all people? To deny the potential benefit to *anyone* on the ground that it won't be available to *everyone* seems illogical. There may be a trickle-down to wider communities over time, but if the research isn't done in the first place then there's no benefit to eventually share. There are many examples of this. For instance, antiretroviral drugs have had a greater impact against AIDS in North America and Europe than in other parts of the world—they have been life-saving, literally, to those who have received them. Should this opportunity be denied to them in the name of equality, just because not everybody can afford them?

Perhaps the question we need to ask is not who *would* benefit, but who *should* benefit from antiaging treatments? In the end, it may be that these advances can't simply be made universal. Perhaps life extension should be permitted only for those who "earn it." This could be defined by the individual contribution, thus extending the life of people that provide some sort of benefit to society. Given the prize at stake, there is an inevitable risk that the system might be open to corruption (imagine a dictator living to be a thousand years old!), but it might provide an interesting alternative to letting the market decide who has enough money to afford it or giving free access to everybody. Can we really afford to let everybody live as long as they wish? What would be the consequences of allowing free access to a therapy that prolongs our lifespan?

The notion of sustainability has had increasing prominence in the past few years. The human population keeps going up. It doubled from 1960 to 2000 and it is estimated that it reached 7 billion on March 12, 2012.[18] It was hoped that it would level off at around 9 billion in 2050 and then slowly start declining. However, studies published in May 2011 by the UN calculate that this number will continue to grow, reaching over 10 billion in 2100.[19] Some argue that the earth is already overpopulated and we are going to be in serious trouble if we don't find a solution soon. Extending our personal shelf life would only exacerbate this problem. However, other experts consider that this is an overstatement and that predictions are being exaggerated. They argue that women nowadays are electing to have half as many babies as their grandmothers, and this decline in birth rate is set to continue.

Either way, the resources of this planet are not infinite. If we want to live longer without exhausting them sooner or later, there are two main options. Either we select just a few people who can have access to the life extension advances or we limit birth to compensate for the increase of elderly citizens. Both approaches would require intervention by the state or even international legislation, which would be deeply controversial.

Population control is not a novel concept. Even in Ancient Greece, luminaries such as Plato and Aristotle extolled the virtues of regulating the size of the population in city-states like Sparta. More recently, advocates of population control have built upon the theories of 18th-century theologian-cum-economist Thomas Malthus (see box: The Reverend's Prophecy).[20]

The Chinese one-child policy is probably the best-known recent example of government-driven legislation on procreation. Introduced in 1978, the policy limits the majority of Chinese families to one child per couple. Although those exceeding this quota are officially fined, there are well-substantiated accounts of abortions and forced sterilization for people breaking the rules. While the one-child policy has succeeded in its primary aim of checking the growth of the Chinese population (it is estimated that there would have been some 400 million more Chinese now if the policy hadn't been implemented), the broader sociological and psychological consequences are still being worked out. Several unintended costs are already clear.[21]

Faced with only one chance to have a child, many Chinese parents have taken action to ensure that they have a baby of their preferred gender. In

---

THE REVEREND'S PROPHECY

Thomas Robert Malthus was a clergyman who lived in Britain in the late 18th and early 19th centuries. He observed that a population that keeps growing in number will eventually fall prey to famine and disease. In 1798, he proposed that unchecked population growth would stop the progress of a society. He also suggested that agricultural advances would one day not be enough to provide food to an ever-growing population. Nowadays these concepts are familiar and broadly accepted in discussions about sustainability. They were, however, very revolutionary at the time he first proposed them.

---

practice this has meant that China now faces a generation with 10–20 percent fewer females than males. In 2005, for instance, an excess of 1.1 million boys were born in China and overall among under 20s in China there were 32 million more males than females. Similar gender imbalances are being seen in India and South Korea, where selective abortion has also been employed. This means, in turn, that there is an excess of men in these countries unable to marry and start a family of their own.

The children that are born in China are, by definition, without brothers and sisters, and many of those children manifest poor social skills of the kind normally honed by interaction with siblings within the home. Some experts have even proposed that societies with an excess of young males are also more violent and unstable (they cite China's crime rate doubling in the past 20 years) and that lack of females increases sex traffic and prostitution. This has also led to an unnatural recasting of the age demographic within China, as four grandparents have had a maximum of two children between them and they in turn have had only one child. There is an aging population, with the number of people over 65 exceeding the number of working-age adults by as much as 6 to 1 in some areas. This clearly makes a nonsense of any traditional pension scheme.

For these practical, as well as ethical reasons, it has been suggested that the Chinese should now abandon the one-child policy. There are signs, in fact, that they will soon be doing so. Already there have been liberalizations such that couples who were themselves both single children have been allowed to have two children. In the Guangdong province the rule was cancelled in 2011,[22] and the Chinese government announced in the spring of the same year that the rule was going to be revised nationwide after the results of the census showed that fertility had significantly decreased in recent years. The fact that those people upon whom the policy was originally subjected in 1978 are now beyond reproductive age themselves enables the government to argue that the experiment has run its course.

It is also important to keep in mind what happened to South Korea, Taiwan, and Singapore, which in the 1960s also subscribed to legislation that encouraged reducing births.[23] In an ironic turn of events, these countries are now facing a worrying decline in fertility (with some of the lower rates in the world) that they are trying to counteract by giving incentives to couples for having two or more children.

Are these real-life examples telling us that it would be practically impossible (and maybe even dangerous) to set worldwide rules to regulate the human population? From a human rights perspective, it is worth noting that Article 16 of the Universal Declaration of Human Rights states that "Men and women of full age, without any limitation due to race, nationality or religion, have the right to marry and to found a family."[24] In other words, in a free country, everybody should be allowed to have descendants and the state should not set out rules limiting this right. There are, of course, ways of "influencing" couples to abide to these rules, for instance, as we said, in the form of economic incentives to those who fulfil the "optimal" quota of children and help demography go up or down. Such strategies have already been applied in many countries, from India to Spain, and could be a way to address this issue.

NEXT STOP . . . IMMORTALITY

We've been talking about extending our lifespan beyond the current maximum of 100-and-something but we haven't discussed what could be the next limit. 150? 200? 500? Once we understand completely why we age, what's to stop us from living forever? Prolonging life would inevitably create all sorts of social and economic problems. But if immortality were achieved, it would require even more fundamental changes in legislation regarding reproduction. Would we only be permitted to have a child if and when an "immortal" died (as a result of accident, exposure to an unexpected disease, or a decision to be "terminated")?

MORE YEARS, MORE WORK

Even if we manage to achieve some sort of balance between births and deaths, a steady increase in older people would have other important effects on society, as we commented above. The key issue is the proportion of "productive" citizens, which are the ones that define the economic potential of a country. As an average, there are now nine working-age adults in the world per each older adult, and this number is likely to go down to four by 2050. The older the population gets, the lower the ratio. If this trend continues, there will not be enough productive people to pay for those that retire.

Currently, mandatory retirement is not allowed in the United States. In other places, like Japan, it's not unusual to continue working after 65. However, in European countries such as Germany, Spain, Greece, and Italy retirement is obligatory at 65. The lowest retirement age in Europe is currently 60, in France, and the highest is in Denmark, 67.[25] These thresholds have remained pretty static for almost a century. However, with mean life expectancy steadily increasing over many years, several countries are already experiencing the need to redraft their legislation on retirement age and pension entitlements—and even the relatively mild proposals are leading to strike action by those affected. In 2010, several countries in Europe, such as Spain and France, started discussing the possibility of extending the current retirement age by a few years. This move sparked heated debates, protests, and strikes, forcing the governments to reconsider. If in the future we have to extend our productive years, it is easy to imagine that some governments will have some serious problems in selling this idea to their citizens.

If aged Western societies are already facing this problem, it's easy to see that a substantial increase in lifespan would only make things worse. If we have to accommodate an exponential increase in old people, then tinkering with the retirement age by the odd year here and there would not be sufficient; we would likely need a hike of several years, possibly decades. Living longer would therefore as a matter of principle mean having to work longer. Some argue that those who remain active for longer are usually also happier and age better, so that shouldn't be much of a problem. But is this what people seeking a longer life are actually looking for?

Although some might be thrilled by the prospect of working until they are 80, many will find this a less attractive proposition. We should also reflect on such developments from the other side of the coin; if older people are hanging on to the top posts in companies for extended periods of time, there could be a detrimental effect on the potential for younger workers to rise through the ranks.

All of this shows that developing life-extending therapies cannot be seen in isolation as a scientific or medical quest—there are wider consequences that need to be thought through before true antiaging treatments hit the market.

THE DEBATE

IN FAVOR:

- Longevity will fulfill an ancient dream of the human race, leading to a happier society.
- It will allow for important people (artists, scientists, politicians, and so forth) to contribute to society for longer. Imagine the benefits if the next Einstein or the next Mozart lived to be 200.
- There's still room on Earth: overpopulation is being controlled by a progressive and generalized fall in fertility, so there's no reason to avoid expanding the lifespan of those who are already here.

AGAINST:

- It is likely that only those with money could afford it, at least initially. It would then create a bigger divide between rich and poor.
- It would eventually exhaust the resources of this planet, unless some serious population control is put in place by governments or access to life extension is limited to a selected few.
- Population control laws could seriously limit human liberties.
- There are important social and psychological impacts to radical changes in the age demographic of populations.
- The system could be abused, and dictators and very rich and powerful people could extend almost indefinitely their control over a population.

CHAPTER 6

Big Brother Is Watching . . . Your Genome

BLOOD TIES

"Tom, it's for you! It's the police."

Caroline Ward had been on her way out of the house to collect the children from school when the doorbell rang. When she answered, she'd been confronted by two police officers asking if her husband was in. She was curious, but she'd left without waiting to find the purpose of their call—if she didn't go now she'd get snarled in traffic and Connor would be late for his soccer practice.

Tom Ward, a freelance computing consultant, saved the document he had been working on and came down the stairs from the converted bedroom that served as his office. He wasn't surprised to have a visit from the police: he had reported his bicycle stolen a couple of weeks earlier and automatically assumed that they had come around to pass on news of their investigation. However, when they invited him to accompany them back to the station, his thoughts skipped to a darker episode from his past. Surely they couldn't be here about that!

It had been nearly 20 years since a fateful October in which Tom had been on a residential course, involving an overnight stay in Birmingham.

He didn't recall what the course had been about, but it was almost certainly some computer programming language or other. Now, as he sat in silence in the back of the police car, his mind began to dredge up long-suppressed memories.

The other delegates on the course had all been considerably older than he was, probably retraining from some other job. He'd felt no affinity with them, so declined their offer of an evening in the hotel bar discussing, he assumed, mortgages and babies and other mundane trivialities of middle-aged life. He chose instead to head out on his own to find a nightclub.

After a couple of false starts—nightclub doormen can be funny about fellows arriving on their own—Tom had ended up at Scarlet Nites. He remembered the name perfectly well. At the club he'd spent much of the evening dancing and drinking with Michelle and her friend Samantha. They'd been getting on famously, and when Samantha started getting frisky with another guy, leaving just the two of them dancing together, Tom was convinced that he'd scored—sex that night was inevitable.

It came as a complete shock to him when he returned from the bar with another drink for Michelle to find that she and Samantha had gone. The other fellow was there, nursing a very clear handprint on his left cheek—it was evident that a significant disagreement had taken place. Tom's surprise rapidly mutated into resentment. He had bought several rounds of drinks at exorbitant prices and now he'd been left with nothing for his troubles. He finished the rum and coke he'd brought for Michelle in a single gulp.

A couple more drinks after that, he finally admitted that the night had been wasted. He left the club himself and started to head back towards the hotel. He was still smarting at the way he'd been let down, when he noticed another girl across the street from him. The ridiculous height of her stiletto heels certainly wasn't helping her progress, but it was evident both from the way she was staggering and her use of the wall for support that she was pretty drunk. Even more than him.

In the days after he'd gotten home from the course, Tom had tried to persuade himself that his initial intention had been to help her. With his sensibilities reduced by alcohol and driven by the unfulfilled expectations from earlier in the evening, he'd simply taken advantage of the circumstances. That's how he justified it to his conscience, at least. The case notes from that night, as he was shortly to discover, described instead a frenzied sexual assault.

When they arrived at the police station, Tom was shown into an interview room. He was surprised at the mixture of emotions he was feeling. Part of him was still hoping that his presence there that afternoon was unrelated to the sordid incident he was now reliving. Strangely, however, part

VOCABULARY

Genes: stretches of "letters" in DNA that contain the information necessary to make proteins, which are the molecules that carry out most of the daily functions inside the cell.

Genetic fingerprinting (or DNA profiling): technique that analyzes certain variable sequences of junk DNA in order to identify a person. Like regular fingerprints, somebody's DNA profile is unique (except in the case of identical twins).

Junk DNA: fragments of DNA that do not have information to produce proteins. They can, however, have important regulatory functions. Compared to the regular genes, certain parts of the junk DNA are quite variable from person to person and can be used in genetic identification techniques.

Low copy number analysis: a relatively new method of genetic fingerprinting (introduced in 1999). It requires much less DNA that other techniques, which makes it very powerful. Its use had been controversial, but now it's accepted in several countries.

Markers: specific regions of DNA that are particularly variable and can be used for genetic fingerprinting. The use of several markers is what gives somebody's DNA profile.

PCR (polymerase chain reaction): technique that allows the amplification of DNA. Starting with a very little amount of DNA, PCR can produce as many copies as needed.

of him had a sense of relief that the truth was about to be uncovered. His confusion was made all the more puzzling by the first question he was asked.

"Do you know Jenny Cartwright?" asked the officer.

Oh God, thought Tom, Jenny's been murdered!

"Yes," he replied with renewed concern. "She's my sister."

"And can you confirm," continued the officer, "that you are her only brother?"

"Yes," answered Tom again. "We have a sister, Frances, but no brothers."

With their relationship confirmed, the officer entered into a fuller expla-nation of the reasons for bringing Tom to the police station.

"Mrs. Cartwright was arrested recently during an altercation at a bar in Twickenham. Although she was later released without charge, a DNA sam-ple was taken from her during the course of the investigation, in accor-dance with standard police procedures. When her details were added to the national DNA database, it was found that her sample matched closely, but not exactly, to a semen sample collected from the victim of a rape carried out in 1993.

"The nature of the correlation between Mrs. Cartwright's DNA and the profile of the assailant in the rape case indicate with a high degree of cer-tainty that the assault was conducted by a first order male relative . . . that is to say a father or a full brother. Our investigations have indicated that your father died in 1988 and as we have just confirmed, you, Thomas Colin Ward, are the only known person fitting the genetic profile of the suspect. I am therefore arresting you on suspicion of sexual assault and rape."

THINK ABOUT IT . . .

Tom's crime was inexcusable, and he deserves whatever punishment that follows. But in a separate incident, his sister Jenny had been released without charge. Despite her innocence her DNA was taken and added to a database.

- Were the police right to take a sample from her?
- Were they right to keep it after she had been exonerated?
- And were they right to use it as evidence against her brother?

THE TRUTH IS INSIDE YOUR CELLS

In many parts of the world, the availability of DNA evidence has revolu-tionized modern policing. Genetic evidence has been used to convict of-fenders, but also to free individuals exonerated when their DNA did not match samples collected from the crime scene. This is what we call **genetic fingerprinting** or **DNA profiling**: identifying a person by the unique sig-nature of their genes. We are familiar with this because we see detectives in movies and TV shows using these techniques all the time, uncovering clues that not even Sherlock Holmes could have guessed, and finally resolving the most perfect crimes. Although it may seem far-fetched, most of what they show in those crime scene TV dramas is indeed possible nowadays . . . albeit the results take rather longer to generate and any one person probably isn't expert in *all* of the methods, as they seem to be in these programs.

The scientific breakthroughs that have led us to this point began at the University of Leicester in the 1980s. In keeping with many important discoveries in science, genetic fingerprinting was actually an accidental by-product of other more fundamental genetic research being carried out by Alec Jeffreys, as he himself is quick to point out.[1] It was an accident that changed his life and that of many others. Jeffreys had been studying the variable regions of the human genome, and in one experiment he was comparing samples from different volunteers, including a few relatives of his technician. At exactly 9:05 a.m. on Monday, September 10, 1984, as he looked at a photo of his latest results, he was suddenly struck that he had uncovered a method to see patterns that were shared only by those with family ties.

The principles behind genetic fingerprinting are deceptively simple. The most important sections of our DNA are called **genes**: stretches of "letters" that contain the information necessary to make proteins, which are the molecules that carry out most of the daily functions inside the cell (see box: The Secrets of DNA). It is currently estimated that there are between 20,000 and 25,000 of these genes in the human genome. This number actually comprises a surprisingly small percentage of our genome, perhaps as low as 1.5 percent. This leaves a large amount of DNA that does not appear to be useful for anything. These have become known rather prejudicially as **junk DNA**, since they seemed to be nothing more than silent remnants of our ancestral development.

As it happens, some sections of this "junk" are now believed to have a very important role in regulating genes.[2] But there are still long stretches of DNA that don't seem to have any specific job. Unlike the sequences within the genes, which tend to be almost identical in all humans, these regions are quite variable. It's this potential for variability that lies at the heart of genetic fingerprinting.[3]

The key thing for identifying individuals is the fact that some of these variable regions could be very different between random people, but don't change very much in any one generation. As a result, it is easy to trace certain regions (that we call **markers**) back to our parents: for each marker, we would get a copy from our father and another from our mother, and certain biochemical techniques would allow us to recognize them. By studying several of these regions simultaneously you could confirm the pattern of familial relationships with great accuracy and build up a combined

THE SECRETS OF DNA

Deoxyribonucleic acid (DNA for short) is a long molecule made up of just four different chemical units (called *bases*), each written millions of times over. These units are the "letters" in which the language of life is written. They are called adenine, cytosine, guanine, and thymine, but are usually referred to simply by their initial letters A, C, G, and T. DNA contains all the information necessary to generate an organism and keep it alive throughout its lifespan.

Different organisms have different amounts of DNA. In humans, our DNA is organized into 46 sections called chromosomes. These are two sets of 23 chromosomes; one set of 23 are inherited from mom, the other 23 from dad. In total, each set of 23 human chromosomes consists of some 3 billion A, C, G, and T letters connected side by side in a very long chain, some of them organized in functional groups that we call *genes*. The order of the letters in the DNA, what we call the *DNA sequence*, is 99.9 percent identical in every human. That's to say, only 0.1 percent of our genome is what makes us different from each other.

pattern that represented a unique signature for one person—the genetic fingerprint.

BUILDING THE ULTIMATE DATABASE

You only need to examine a surprisingly small number of these genetic markers to obtain a combination that is specific to each individual. This simplicity allows us to set up databases of genetic profiles that can accumulate data from millions of people without requiring huge amounts of digital storage. This is a modern version of the traditional fingerprint archive, and it has the advantage of being easily accessible with a computer. A new sample obtained from a crime scene can rapidly be compared to millions of stored profiles, which makes the identification of a suspect much easier and faster than before. The catch, of course, is that his (or her) profile has to already be in the system to be recognized.

For this reason some countries have started to take DNA samples from criminals and offenders and store their genetic profiles in centralized databases (see box: Profiles and Samples).[4] For the National DNA Database

PROFILES AND SAMPLES

It is important to draw a clear distinction between a DNA *profile* and a DNA *sample*. A DNA profile is a series of numbers that contain the information for different genetic markers. This string of numbers can easily be stored in a computer and then compared automatically with the numbers for anyone in a database. A DNA sample, on the other hand, is the actual biological material collected from a cheek swab, blood, semen, etc. A sample carries much more information than the specific DNA profile: it contains the full genome of a person.

(NDNAD) in England and Wales, a standard set of 10 markers is used, plus an additional sequence on the sex chromosomes. For the American Combined DNA Index System (CODIS), 13 markers and a sex identifier are used.[5] Even more importantly, thanks to current techniques, such as the **polymerase chain reaction (PCR)**—a clever way to make lots of extra copies of specified pieces of DNA—we only need a very small amount of DNA to obtain this information. This means that a bit of saliva, a few hairs, or a drop of sperm could be more than enough to get somebody's genetic fingerprint.

As we will discuss below, the use of DNA databases is controversial, and many countries have decided that, for example, the infringement of civil liberties does not justify their routine use.[6] It is beyond dispute, however, that they have been extremely useful in solving all kinds of criminal cases, even some that had been committed decades before, like the one we described at the beginning of this chapter.

YOUR SINS WILL FIND YOU OUT

Tom's story is fictional, but it has echoes of many real life cases. Sometimes a DNA sample taken from a person for one alleged crime will bring up a match to a sample collected at a different crime scene. For instance, in 2006, a fight broke out in a pub in Surrey, in southern England. The police took DNA samples from several people present that night, including the pub's chef, Mark Dixie. When his DNA profile was added to the U.K. National DNA Database, it was found to be the same as the profile of a suspect wanted for a murder committed in London nine months earlier.[7]

With the aid of DNA technology it is sometimes also possible to go back and reinvestigate crimes committed long before the concept of a DNA profile was ever considered. In one British case, Christopher Smith was shown to have been responsible for a murder committed 35 years earlier. Smith was arrested in 2008 for drunk driving, and a DNA sample was taken. He died of a terminal illness a matter of days after his arrest for this offence, but his DNA profile was retained. More recently, police applied the latest tools of forensic DNA work to samples relating to the murder of Joan Harrison, which took place in 1975. There was a match to Christopher Smith's profile. This new evidence motivated the police to look more closely into the case. They uncovered sufficient further proof of his involvement for prosecutors to declare that, if Smith had still been alive, they would have charged him with the murder.[8]

CAUGHT BY THE WRONGS OF YOUR FAMILY

Just as Tom was to discover in our scenario, several real-life examples have shown that it can be the actions of a relative that leads to the police knocking on your door. In 2003, truck driver Micky Little was killed while driving along a British motorway. As he passed under a bridge, someone threw a brick through the window of his truck cab, which caused him to have a heart attack. The police found traces of DNA on the brick and a matching sample in a stolen car abandoned nearby.

The person who threw the brick did not have a criminal record at the time, so their DNA was not in the database. However, there was a partial match to someone else already in the system, and that person was highly likely to be a family member of the brick-thrower. With this as a starting point, the police began to look for a relative of the person whose DNA was already on file.[9]

They could get some information directly from the DNA on the brick. The presence of a Y chromosome in the DNA told them that the suspect was male (women have two X chromosomes, while males have an X and a Y). Other genetic characteristics suggested he was very likely to be white skinned (see box: A Genetic Picture).[10] They then added some nongenetic hunches about the suspect. Looking at the type of offence and the location where it happened, the police also speculated that they were looking for someone aged under 35 who lived close to the crime scene.

When they combined these various factors, shop assistant Craig Harman came out as the most likely perpetrator. This was confirmed by a sample taken directly from Harman. At a trial in April 2004, he pleaded guilty to manslaughter and was sentenced to six years in prison. He was therefore

A GENETIC PICTURE

For a number of years, forensic scientists have been excited by the possibility of being able to make predictions about a suspect's physical features, known as their external visible characteristics, based solely on any genetic evidence they may have left behind at the scene.

This approach is now becoming a reality. There is no need for the suspect's DNA to be in any existing database—aspects of their appearance can be inferred directly from their genetics. Predictions of eye color, hair color, skin color, and even age (within a range of about nine years) are becoming possible. Broad distinctions of race (for instance between European, African, sub-Saharan African, or East Asian) are already a reality.

These techniques, termed *forensic DNA phenotyping*, only provide statistical probabilities, not certainties, and they can also be seen as an invasion of the suspect's privacy. For these reasons, they are currently forbidden in countries such as Denmark, Belgium, Germany, and some states in the United States.

the first person in the world convicted on the basis of a "familial search" of DNA evidence.[11]

The case of Keith Davidson is another recent example. Police had long since abandoned hope of solving a 1990 rape case, for which they had no suspect. In 2004 they re-examined the forensic samples using improved DNA methods and managed to generate a DNA profile for the attacker, but nothing on the database matched. Then in 2006, Keith Davidson's daughter was cautioned for assault and her DNA was taken. It proved to be a partial match to the 1990 case, and a subsequent sample taken from Davidson himself was shown to be a perfect match. He was sent to prison for eight years.[12]

DNA EVIDENCE CAN ALSO SET YOU FREE

Interestingly, the very first criminal use of genetic fingerprinting resulted in the *release* of the police's prime suspect. Two 15-year-old schoolgirls in neighboring Leicestershire villages, Lynda Mann and Dawn Ashworth, had been raped and murdered in nearly identical ways three years apart (in 1983 and 1986, respectively). Following the murder of the two local girls, Leicestershire police turned to Alec Jeffreys to use his new-fangled DNA test. It showed that the man they had in custody was in fact innocent. Richard Buckland, the prime suspect, was categorically *not* responsible, despite the fact that he had confessed to the first murder. Fortunately the police had the good sense to believe the DNA test rather than assume it must be wrong.[13]

If Buckland hadn't killed the two schoolgirls, who did? In a turn of events worthy of a Hollywood script, the police decided to retain their confidence in the emerging DNA technology. Convinced that the killer was a local man, they decided to conduct a voluntary screen of men in nearby villages. Despite achieving a 98 percent turnout, the DNA dragnet seemed to have failed. But then, a woman in her local pub overheard bakery worker Ian Kelly boasting that he had pretended to be his colleague Colin Pitchfork to cover for him when it was time to give a sample, and that he had got £200 as a reward for the favor. Pitchfork had told Kelly that he couldn't do it as he'd already give a sample posing as another friend who needed an alibi. Following the woman's tip-off to the police, Pitchfork was arrested and tested; his DNA was a perfect match to the murderer. He was sentenced to a minimum of 30 years in jail.

Since 1989, more than 250 people in the United States alone have been exonerated as a result of DNA tests carried out after they had been convicted and jailed.[14] Take the case of Kirk Bloodsworth, who had been on death row for two years before he finally persuaded the authorities to conduct the DNA test that proved he had not carried out the brutal child murder for which he had been convicted. He was released and pardoned in 1993. And in December 2009, James Bain was released from a Florida jail when DNA evidence showed that he was not guilty of raping a child. Bain had served 35 years in prison before genetic evidence confirmed his innocence.

SO DNA DATABASES ARE A GOOD THING, RIGHT?

The cases discussed are just examples of an ever-increasing catalogue of investigations where DNA evidence has been crucial. Criminals who might otherwise have gotten away without punishment are made to account for their actions. People imprisoned for crimes they did not commit are freed. On the face of it, the development of DNA profiling and the establishment of databases seem to be a "win-win" situation. There are, however, a number of serious issues associated with both the use of DNA evidence in general, and with particular ways in which people's DNA has come to be included in some of the larger databases. We'll discuss some of them.

For more than 15 years, police in Germany were on the trail of a mysterious female murderer. This woman was so notorious she had two nicknames: the "Phantom of Heilbronn" (the name of a German city where a young policewoman was the victim) and the "Woman Without a Face" because although her DNA had been found at over 30 crime scenes across Germany and neighboring countries Austria and France, no one knew what she looked like. When an eyewitness described seeing a suspicious man near one of the crime scenes, there was even speculation that perhaps this woman was a transsexual: genetically female, but living with the persona of a man.

Then in 2009 the case of the "Phantom" came to a dramatic conclusion.[15] The woman whose DNA had been linked to so many crimes across that part of Europe was, in fact, a Bavarian worker in the factory that makes the swabs used to collect DNA samples. Despite the fact that conditions were supposed to be sterile, minute traces of the woman's DNA (perhaps in the form of sweat, saliva, or skin cells) was getting onto the swabs. Her genetic material was being taken *to* the crime scenes, not collected *from* them.

The story of the "Woman Without a Face" is a particularly stark example of a general concern, namely, the ease with which a crime sample can be contaminated with the DNA of someone else. DNA belonging to an innocent person might be found at the scene because they really had been to that place, but at a different time, days or even months before a crime took place there. If there are mistakes in the handling of the DNA sample after collection, there is also the risk that it might be contaminated with someone else's genetic material at that stage. You really don't need much for this to happen (performing standard tests only requires DNA from a few dozen cells). The potential for cross-contamination with DNA from a different crime scene might leave somebody guilty of a minor offence facing apparently irrefutable evidence that they perpetrated a much more serious crime as well.

We also have to consider that it is relatively easy to plant evidence in a crime scene by secretly collecting a DNA sample from the person to be framed. Just a few hairs or skin flakes can fool the police into thinking that somebody had been in a certain place. Without the proper controls, genetic profiling could turn society into a nightmare in which an innocent person can be accused of practically any crime (see box: Fakin' It).[16]

The relevance of contamination and foul play has become more significant with the introduction of ever more powerful ways to build up a DNA profile from smaller and smaller amounts of starting material. For instance, even some enthusiasts who are supporters of genetic fingerprinting as a forensic tool are nevertheless worried that a technique called **low copy number (LCN) analysis** may tarnish the reputation of DNA evidence. With this method, only a few cells are sufficient to extract the information needed to

---

FAKIN' IT

In 2010, Israeli scientists reported that DNA evidence could be fabricated. They amplified tiny amounts of DNA in the lab and mixed it with someone else's blood, from which all white cells had been removed (our red blood cells don't contain DNA). The result was a "fake" blood sample that fooled even the experts. They were also able to create from scratch a sample that contained the markers usually found in a database for a given person. With the amount of DNA found in a hair, for instance, we could generate liters of fake blood genetically identical to that of a suspect.

These techniques needed to plant false biological evidence in a crime scene are fairly simple and could be replicated easily by somebody with the right biological knowledge. This could cast a long shadow over the reliability of DNA tests, which are nowadays critical to an important percentage of cases. To address this, the authors of the study set up a company that sells tests to help differentiate real samples from fake ones.

---

build a genetic fingerprint. Therefore, it is much more efficient, and this has allowed the police to reopen cases that had been closed for years due to the fact that the samples obtained from crime scenes were not good enough. Obviously, the problem is that this method is also more susceptible to errors. Any contamination, even insignificant, can be amplified and mask the results. This could easily lead to an innocent person being imprisoned. Experts still don't agree on whether benefits justify the risks. For this reason, low copy number analysis has only been applied in a few countries, such as the United Kingdom, the United States, the Netherlands, and New Zealand.[17]

Confirmation that these concerns may be justified came in 2007 when a judge dismissed the case against Sean Hoey. Mr. Hoey was accused of involvement in the 1998 bombing at Omagh, Northern Ireland, in which 29 people were killed. A major plank of the case against him rested on DNA evidence found on parts of the bomb. The low copy number technique had been used to generate the DNA profile, and the judge questioned the robustness and quality control of the approach.[18] Because of this, the technique was banned in the United Kingdom for two months, while a committee was investigating its accuracy. It is now being used again following more strict protocols. In 2009 a judge in California ruled that it could not be used in court because it was still not accepted by all experts, but in 2010, a New York judge ruled that the technique was reliable enough. Clearly the controversy has yet to be resolved.[19]

PROTECTING THE INNOCENT?

Alongside concerns about accidental or intentional contamination of crime scene samples there are also worries about the scope of the databases, and about the sort of material that is being stored.

The U.K. National DNA Database was first established in 1995. Originally samples could only be taken from suspects believed to have committed a serious offence, and only with their consent. In the intervening period a series of changes in the law have seen the need for consent waived, and the grounds for taking and storing of DNA information broadened to include arrest for any offence, even if you are never actually charged or are charged but are subsequently found not guilty. DNA from children as young as 10 can be included.

Such has been the impact to these changes that by 2010 there were about 5 million individuals profiled in the British database,[20] many of whom were innocent people who had never been found guilty of a crime. This figure represents nearly 10 percent of the overall population, the highest percentage in any country. This surprisingly large average actually masks distinct variations between different subgroups. For example, more than 30 percent of black males over the age of 10 have a profile in the database. The trend towards inclusion of DNA profiles from more of the population for increasingly tenuous reasons had also been echoed in America, where the FBI's CODIS database now has over 10 million profiles.[21]

This has raised some serious ethical concerns. Is it fair to store data from people that have never committed a crime? Critics argue that the intention to protect the public from criminals has sacrificed protection of the individual freedoms of the innocent. In a recent test case, the European Court of Human Rights decided unanimously that the U.K. government had violated the applicants' right for respect of their private and family life (see box: *S and Marper v. the United Kingdom*).[22]

One of the issues involved in storing the DNA profiles of only those who committed or have been involved in any manner with a crime, even a minor one, is discrimination. They would have the disadvantage of being much easier to catch if they break the law again, since their information would be in the bank. This is a serious disadvantage when compared to those who have never been caught. Some may argue that the "usual suspects" deserve a tougher response and that it's all right to make it more

S AND MARPER V. THE UNITED KINGDOM

The first party to this lawsuit was an 11-year-old child arrested for attempted robbery (hence he is known simply as S to protect his identity). At the time of his arrest, his fingerprints and a DNA sample were taken. In June 2001, he was found not guilty, but police refused to remove his details from the National DNA Database.

Michael Marper was arrested in 2001 and charged with harassing his partner. He too had his fingerprints recorded and a DNA sample was taken. The case was officially discontinued when he and his partner were reconciled. Here too, however, the police refused to remove his profile from the bank.

In December 2008 the European Court of Human Rights decided that keeping of DNA profiles and samples even after a case against someone was dropped represented a breach of Article 8 of the European Convention on Human Rights (echoed also in the U.K. Human Rights Act), the right to respect for their private life. As a consequence, the U.K. government now is supposed to be limiting the time that genetic information can be held if the person they refer to is found to be innocent. At the time of this writing, however, very little practical change has been introduced.

difficult for them to continue to pursue a life of crime. But they are not the only ones being discriminated against. Keep in mind that some socioeconomic backgrounds are more prone to be interrogated by the police (as we said above, there are up to three times more black men in the databases than from other ethnic groups). Thanks to this, a criminal is more likely to be caught if he's black or if he comes from a poor neighborhood. Not everyone has the same chances of escaping justice, and this could be seen as an attack against their rights.

Some have suggested we should move towards a situation in which everyone will be profiled at birth and databases will contain information about every single citizen.[23] They don't see this as different from current databases in some countries, which include fingerprints and unique identification numbers. This would certainly make identification of suspects much easier, but how could we protect, say, the innocent opponents of a totalitarian government from having their data planted at a crime scene? An agency with access to the genetic samples could, in principle, frame anybody for a crime, thus "legally" disposing of any citizen they chose.

This leads to another area of concern: the storage of the actual DNA samples, not only the DNA profiles. Remember that a DNA sample has all of our genetic information, not just the few selected markers found in DNA profiles. In the hands of the unscrupulous it could tell people about our susceptibility to different illnesses, and it could perhaps divulge that someone's "dad" is not genetically their father after all. And if we push the limits of what could be possible in the future, this material could even be used to clone somebody unbeknownst to him, as we discussed in Chapter 2. Campaigners therefore argue that the samples themselves should be destroyed in order to avoid misuse of the wider information.

It is hard to dispute that DNA evidence is proving to be an excellent tool for police work. When used appropriately, DNA profiling can help to exonerate the innocent and to catch the guilty. However, the consensus now is that governments and their police forces need to exercise greater care to protect the autonomy and privacy of individuals, including the destruction of all DNA samples once a DNA profile has been developed from the material, and the protection of the information stored. They also need to make it easier for innocent people to have their profile removed from the DNA database.

THE DEBATE

IN FAVOR:

- Having a large database of genetic profiles allows us to identify more criminals, not only for current cases, but also for reinvestigating cold ones.
- Crimes are more likely to be reoffenses by known criminals rather than new offenders, so it makes sense to have their genetic details on record.

AGAINST:

- Genetic information of innocent people is stored without consent. This is an attack against their privacy.
- Current policies discriminate against certain social and ethnic groups.
- Mistakes in handling DNA evidence could lead to innocent people being found guilty of offences.
- Potential misuse of genetic information by the government or criminals with access to it.

CHAPTER 7

Something on Your Mind?

ALL IN A NIGHT'S WORK

As Luis slipped from the alleyway into the Madrid night, he wore the smile of a man satisfied with a good night's work. In the backpack slung over his shoulder he had a latest model iPhone (first time he's bagged one of those), a laptop, and a PlayStation. As he played over the layout of the apartment in his head, he was pretty sure he hadn't missed anything that was easy to carry—the flat-screen television was too cumbersome and previous experience had told him that taking DVD players was no longer worth the effort. The painting on the wall may have been valuable, but he didn't know enough about art to take the risk.

All in all, Luis was pleased with a well-executed burglary. He had, as always, been meticulously careful to leave no evidence to help the police. He had checked the building for security cameras a few days in advance and had worn gloves and a mask throughout. He'd sell most of the stuff right away using his regular channels, and tomorrow night he would be rich enough to go out and celebrate.

The following day, Luis met with Silvino, his regular fence. Silvino didn't always offer the best price for the "previously owned" items that Luis had on offer, but he was willing to deal in a variety of goods, which made him a useful contact. They settled on €150 for the laptop and the PlayStation. Luis decided not to mention the iPhone—he'd needed a new phone for a while and this might be just the ticket. There would always be the possibility to sell it to Silvino later.

With some money in his pocket, he went home to get ready for the night out he had promised himself. He'd invited Sofia to go to dinner with him—she was going to meet him at his apartment at eight. Luis had a shower and, having settled on an outfit, started to examine the iPhone. It had a passcode, but this turned out to be 1234 (as usual . . . why were people so predictable?). He browsed through the music library. Unfashionable stuff, "old folk's songs" as far as Luis was concerned, not his taste at all. Nothing he would even recognize.

Shortly before eight, there was a knock at the door. Expecting Sofia, he didn't bother to check through the peephole first. That was his mistake, as he discovered when he found himself face-to-face with two policemen.

"I didn't do anything!" he complained automatically.

One of the agents was carrying an open laptop. He pushed Luis away from the door and got into his apartment without saying a word. He went straight to the iPhone, which was now on the kitchen table. Then he smiled.

"Tell that to the judge! Looks like we caught you red-handed, pal!"

A few hours later, a lawyer was explaining to Luis that thanks to an app called FindMyiPhone and the phone's GPS chip, it was easy to locate any unit that had been previously registered on Apple's Web site. If Luis had been more savvy, he'd have known about this feature and steered well clear! Too late now, the lawyer told him.

Things didn't look good for him. The lawyer insisted that the best thing was to cooperate, but Luis was planning to stick to his story: he had no idea how that phone had gotten into his apartment. Probably one of his friends left it there. He had had a big party the night before (that was his alibi, and it would be easy for him to find people willing to back it up); any of the guests could have been the burglar, just not him.

Luis was surprised when the police, after going over his declaration, told him they were transferring him to the hospital.

"But I'm not ill! Why are we going to the hospital?" he asked his attorney.

"I gather they have a new interview procedure they want to try that they can't do here," replied the lawyer.

"Going to give me a truth drug?" Luis joked to the senior policeman.

"Closer than you think," was the unexpected reply as the officer collected the papers from the table and invited the assembled group to make their way out to the waiting cars.

Once at the hospital, Luis was shepherded towards the radiology department and into a room dominated by a machine that looked like a giant

VOCABULARY

CAT scan (computerized assisted tomography): imaging technique in which a computer assembles different images obtained by exposure to X-rays, in order to reconstruct the three-dimensional image of a part of the body.

Electroencephalography (EEG): uses a network of electrodes connected to the outside of the head in order to monitor changes in the electrical activity going on inside the brain.

fMRI (functional MRI): **MRI** that measures the change of blood flow in each part of the brain, in order to determine which neurons are active in any given situation.

Magnetic Resonance Imaging (MRI): an imaging technique that does not use radiation but instead measures magnetic resonance of certain atoms inside cells to assemble a picture with the aid of a computer.

Positron emission tomography (PET): imaging technique that follows the distribution in the body of a tracer that emits positrons. Depending on the type of tracer, we can detect the activity of different tissues. This, combined with a regular CAT scan, can mix information about image and function, similar to what an fMRI would do.

Quantitative electroencephalogram (QEEG): detects regions of abnormal electrical activity in the brain and is more frequently used to study the brains of people with dementia or injuries.

cotton reel, a structure that was clearly the reason both for this room's existence and for their journey from the police station.

A woman in a white coat rose from behind a computer in the corner of the room.

"Hello Mr. Gómez," said the technician, "and welcome to the new crime-scene recognition service. This machine is a **magnetic resonance imaging** scanner, or MRI for short. In a few moments we are going to ask you to empty your pockets of anything metallic and then to lie on this couch." She gestured towards a bed-like platform level with the hole through the giant cotton reel. "When everyone's ready, you will be moved so that your head is inside the machine. It is a little claustrophobic within the chamber, and slightly noisy when it is running, but you won't feel anything.

"The scanner will be monitoring activity in different areas of your brain. We will give you instructions via an integrated intercom and you will be shown a series of images on a screen within the chamber—that's it really. The test will be over within about 20 minutes. Do you have any questions?"

Luis, still thinking of the experience as more exciting than threatening, was nonetheless prompted to ask the obvious one.

"What exactly are you expecting to get out of this?"

"The test involves showing you photographs of various locations," replied the technician. "If you recognize any of them, your brain will automatically respond in a different way when compared with an image of somewhere you don't know. If everything is okay, shall we get started?"

He was invited to lie on the mechanized platform and the technician positioned his head securely into a helmet-like frame made of plastic. Once they were confident that he was okay, they slid Luis forward inside the scanner.

"Can you hear me?" her voice asked via the intercom, "If you can, just say 'yes' aloud. There's a microphone inside the chamber that will pick up your reply."

"Yes," said Luis.

"Okay, first off, we just need to check that the machine is working properly. Can you look at the screen, Luis, and tell me if you know the place shown in the first image."

On the screen, a photograph of a cathedral with eight tree-like spires came into view.

"It's the Sagrada Familia in Barcelona," said Luis.

"That's great. What about this next one?" asked the technician.

The image of Gaudí's famous church was replaced by a photo of trees in an orchard. It looked idyllic, but not familiar.

"No," replied Luis.

"That's good too," said the technician. "You're not supposed to know that one. It's the lemon grove in my grandmother's garden. Let's try another."

The third image was in stark contrast to its predecessor; a ransacked room. Judging by the desk and the filing cabinet—the contents of which were now strewn across the floor—it was somebody's home office after a visit from an uninvited guest.

"Never been there," offered Luis.

All of a sudden, his bravado began to appear misplaced and he actually started to feel a little nervous. What if this machine really *could* read his mind? That mess wasn't his style of burglary, but he was suddenly concerned that his brain might betray him when they came to a crime scene that did bear his hallmarks.

Over the next few minutes, Luis was shown a succession of similar photos. Each time, he was able truthfully to deny any knowledge because he really had not seen the rooms before. Just as he was starting to think that the police actually didn't have anything that they could match to him, he received a rude awakening. On the screen he saw a photo of an apartment that was without doubt familiar. The flat-screen TV, the painting on the wall . . . it was the apartment he had raided yesterday, the very place where he had stolen the iPhone that had gotten him into this position.

"No, don't know that one either," he said, but inside his heart was pounding.

Did they believe him or did their new technology render his denial useless? Had his brain involuntarily confessed?

They showed Luis images of three more apartments, one of which brought no flicker of recognition, but two of them did ring bells; he recognized the tropical fish tank in one and nude portrait in the other. Again he denied any knowledge, but he wasn't sure he'd managed to sound convincing, and he was increasingly fearful that his mind had told a different story with rather more clarity. At this point the technician's voice explained that the test was over, and that she would help him out of the scanner.

From inside the MRI machine, Luis had not been able to make out the conversation taking place between her and the policemen. However as the platform on which he was lying was pulled from the machine, he could sense a general buzz of excitement in the room. Luis was invited by Sergeant Diaz to stand up, and as he did so he heard the sergeant continue:

"Luis Gómez, I am arresting you for three counts of burglary. You have the right to remain silent . . ."

THINK ABOUT IT . . .

Is it fair to require suspects to take a test of this kind? Can we force people to "show" us what's inside their minds?

Could this technique be misused in other kind of interrogation, maybe illegally?

Can these results be considered reliable? Could they ever be used in court?

PEEKING INSIDE YOUR HEAD

The mind plays such an important role in our notion of humanity that we should not be surprised at the interest scientists have shown in dissecting how it functions. The fact that the brain is housed inside a hard cranium specifically designed to protect it does, of course, present a fundamental difficulty when it comes to studying how it works. Attempts to map out the link between the structure of the brain and its various functions have often involved anatomists cutting through the skull in order to get at the tissue inside. This is clearly an invasive and risky procedure, and over the centuries people have therefore looked for various ways to map the brain without having to carry out such literal dissection (see box: The Shape of Things).[1]

A key breakthrough came in 1895 when German physicist Wilhelm Röntgen discovered X-rays. His invention offered a way to take "pictures" of what's inside our bodies without having to use a scalpel to open them. In the intervening years, medical imaging techniques have been improving and diversifying at an amazing speed. Some of these methods allow us to make structural observations about the brain. Modern-day scientists now have an impressive range of more reliable tools for studying the brain. Computed axial tomography (CAT) scans, for example, involve taking a series of X-ray images on the brain and is useful for showing where injuries have occurred or to spot tumors.

Other approaches go even further and are able to show the brain at work. For instance **electroencephalography (EEG)** uses a network of electrodes connected to the outside of the head in order to monitor changes in the electrical activity going on inside the brain. EEG has been especially useful in studying sleep patterns. **Positron emission tomography (PET)**

THE SHAPE OF THINGS

One of the most well known, but now discredited, approaches to understand how our brain works was phrenology. Popular in the mid-19th century, phrenologists held that the shape of the skull revealed the relative size and importance of different sections of the brain responsible for controlling characteristics such as self-esteem, friendship, combativeness, and musicality. By studying the contours of the skull it was believed that you would have clues to the patient's underlying personality and even intelligence.

studies brain activity differently, by looking at the way special radioactively labeled nutrients are used by the brain. But the technique currently raising the most interest, and the one used to examine Luis in our scenario, is called **functional magnetic resonance imaging, or fMRI**.

The physical chemistry involved in fMRI is quite complex, and we do not need to discuss it here. All we need to know is that the method allows us to monitor the flow of oxygen to different areas of the brain as a proxy for showing where activity is taking place.[2] Some scientists are using fMRI to study the link between parts of the brain and aspects of our personality or emotions. At the moment, no criminal justice systems are actually using fMRI for crime scene recognition in the way that our story suggested. But before burglars like Luis relax too much, there are certainly some who are considering it. Besides, other uses of brain imaging techniques have already made it into the courtroom, as we'll consider in a moment.

The machines needed to do an fMRI scan are not yet portable and require the full collaboration of the subject, who has to remain still during the assessment, otherwise the pictures get blurred. Thus, we can't consider this an invasive technique. Yet. As technology advances it may be possible to "see" what goes on in our minds without having to ask permission. It will never be equal to "mind reading" as we understand it, but nevertheless it may become possible to see things that you would prefer to keep hidden from view. For instance, there have been suggestions that sexual preference, capacity for teamwork, or racism can be detected with these techniques, which could maybe even predict personality and intelligence.[3] This could be used as a diagnostic tool in some situations, but it could also have more controversial applications, such as screening a potential employee or interrogating a prisoner. Because of this, a series of ethical rules may have to be put in place to protect the innocent. Exploitation of brain imaging for targeted advertising is also being considered (see box: Your Ad Here).

IS THERE ANYBODY IN THERE?

The most exciting uses of fMRI are actually in the determination of consciousness. People who have suffered traumatic brain injury can end up in a range of coma-like states. Depending upon the exact symptoms, the diagnosis might be that the patient is suffering locked-in syndrome, in which they can feel but can't communicate (see box: Locked Inside Their Bodies), or is perhaps in a persistent vegetative state, in which all voluntary cerebral activity has ceased (doctors sometimes talk about them being "awake but not aware"). But how can we tell if these patients really know what happens around them?

There was no answer to that question before some ground-breaking research done by a team of scientists in England and Belgium.[4] Initially, the researchers asked healthy people to think about carrying out various tasks (without actually doing them) and monitored their brains to see which areas were active. They then developed a system by which two tasks that caused clearly distinguishable regions of the brain to light up became proxies for "yes" and "no" answers. They actually asked people to think about playing tennis if the answer to the question was "yes" and to think about walking through their house if the answer was "no." Having established that the system worked for healthy volunteers, they tried it with patients in various coma-like states. Amazingly, a subset of these patients was able to understand the instructions and to answer yes/no questions using this process. This offers a means of communication with the patient inside their motionless body and shows, at least for some patients, that they are far more awake than had previously been acknowledged.

This discovery has fueled another longstanding ethical debate: the right to choose death. Patients in persistent vegetative state, deemed to have no hope of recovery, have been allowed to die by withdrawal of feeding (like the cases of Tony Bland in the United Kingdom and Terri Schiavo in the United States).[5] The decision was made for them in absence of any sign of brain activity. Is that acceptable given that some people who were thought to be similarly unaware have now been shown to have greater consciousness than was believed? Do relatives have the right to make that decision? If we use functional brain imaging to solve this challenge, we may find ourselves facing a bigger problem. What if someone being kept alive in a coma-like state, able to make contact via the brain scanning route

LOCKED INSIDE THEIR BODIES

The autobiography and subsequent film *The Diving Bell and the Butterfly* tells the moving story of French journalist Jean-Dominique Bauby. In December 1995 Mr. Bauby suffered a stroke and slipped into a coma. He remained in this state for more than a year and a half, at which point he became mentally aware of the environment around him, but was unable to move any part of his body except for very limited movement of his eyes. This condition is known as locked-in syndrome. Amazingly, Bauby was able to dictate his memoir by blinking in response to a colleague reading a list of letters. This laborious process took about 10 months. He died before the release of the film.

Another victim, Martin Pistorius, whose autobiography *Ghost Boy* was published in 2011 was more lucky . . . eventually. When he was 12, he slipped into a coma and ended up in a care center. It took 10 years for somebody to realize that, although he couldn't move, he was aware of everything happening around him. He then started communicating through a computer and eventually recovered some of his motor functions. In his book he recounts the abuse he had to endure in the hands of nurses that thought he couldn't hear or feel anything. He's now married and runs a business from his home in England.

and answer yes/no questions, gives a clear indication that they wish to be allowed to die? Under those circumstances, do family and doctors have a duty to accede to their wishes?

THE LEGAL USES OF BRAIN IMAGING

There are certainly people within the legal system showing interest in the use of fMRI. Companies like No Lie MRI and Cephos Corporation are now offering the technology as a more accurate lie detector service than traditional polygraph testing (which is based on pulse rate, sweating, and other physiological signs). Their Web sites claim their service offers "the first and only direct measure of truth verification and lie detection in human history."[6]

Not everybody agrees that this is accurate. Although fMRI-derived data on personality has been used in court (see box: Twisted Brains),[7] an American judge ruled in June 2010 that fMRI was invalid as a lie detector because

TWISTED BRAINS

fMRI is still fairly new, and it doesn't have the widespread scientific approval that other well-tested techniques (like MRI itself) have. The first courtroom use was actually in Chicago in 2009, on Brian Dugan, a confessed killer and rapist of a 10-year-old girl, 7-year-old girl, and a 27-year-old woman. Neuroscientist Kent Kiehl was studying the brain of psychopaths with fMRI and Dugan's lawyers invited him to analyze their client. His data was later presented to the jury as an example of a brain that was not functioning normally, and therefore, it was argued, he was not completely responsible for his actions. Nevertheless, Dugan was sentenced to death.

"the technology is unreliable and has not been accepted by the scientific community."[8]

Concern has been raised that too much reliance has been placed on the accuracy of tests. The areas involved in control are small in size and, despite people's brains having similar architecture, there will be important differences in relative proportion of different sections (rather like general body shape, in fact). It is therefore possible to misinterpret the results. In other words, the actual errors rates of this technique are still being disputed and it is therefore unlikely that their use will become widespread soon.[9]

There is also interest in using various brain imaging techniques to determine whether a suspect knew their actions were wrong, particularly if they are a child or someone with learning difficulties. Death row inmates in American jails have already used this approach to try to earn a stay of execution (like in Brian Dugan's case). One of the earliest known cases is that of John Hinckley Jr., who shot Ronald Reagan in 1981. His lawyers used a **CAT scan** as evidence that he had suffered brain shrinkage and argued that this indicated a mental defect. Other experts involved in the trial considered the scan to be normal, but Hinckley was eventually found not guilty by reason of insanity.[10]

Another technique known as "brain fingerprinting" has received significant interest from both the FBI and the CIA. A version of electroencephalography, brain fingerprinting relies on subtle difference in the electrical signals within the brain depending on whether certain information is known or not known. It does not work as a lie detector, nor does it reveal how the information got there, only that it is a detail about which the test subject is now aware. Anyone wanting to use this as a method of discerning guilt or innocence therefore needs to make sure that the answers for which they are probing are not publically available.

The two most celebrated uses of brain fingerprinting involve Terry Harrington and James Grinder. Harrington had served 25 years in prison for murder before brain fingerprinting was used to overturn his conviction. Grinder was being investigated for the rape and murder of Julie Helton. As part of the gathering of evidence against him, a brain fingerprint was conducted. This seemed to suggest that Grinder knew details of the case that only the perpetrator would have known. This result is considered pivotal in Grinder's decision to confess and offer a guilty plea.[11]

In December 2005, Peter Braunstein was arrested in Memphis, Tennessee. He had been on the run for six weeks since the Halloween night imprisonment of a former colleague in her New York apartment, followed by several hours of sexual assault. At his trial in 2007, Braunstein did not deny that he had carried out the attack, but argued instead that he was incapable of having planned such a crime. To support this case, Braunstein's defense team offered PET scans of his brain and an expert willing to testify that the results showed Braunstein had impaired functioning of the areas of his brain involved in moral judgement such as the initiation or cessation of certain behaviors. In this case, however, the jury were unimpressed by this apparent scientific excuse for his behavior. In their minds, the fact that Braunstein turned up at his former colleague's apartment block dressed as a fireman and that he threw a smoke grenade in the lobby to simulate a fire in order to get her to open her door looked pretty much like a premeditated crime. They took just four hours to find him guilty and he is currently serving a minimum of 18 years in prison.[12]

In a separate 2005 case, Grady Nelson stabbed his wife 61 times and then stabbed and raped her 11-year-old daughter, who was mentally handicapped. The attorney in Florida sought the death penalty for murder, but the jury opted for life in prison instead. The key evidence presented by the defense was obtained using another form of EEG called **quantitative electroencephalogram (QEEG)**, which detects regions of abnormal electrical activity in the brain and is more frequently used to study the brains of people with dementia or injuries. The QEEG test showed Nelson had some unusual activity in the left frontal lobe of his brain. Neuroscientist Robert Thatcher convincingly argued that this could be an explanation for Nelson's bizarre crime, since damage to the frontal lobes can leave people with difficulties in controlling their behavior and understanding the consequences of their actions. It was suggested that these anomalies in his brain may have stemmed from documented head traumas Nelson had previously suffered.[13] As with fMRI, however, there are experts who believe that QEEG is not quite ready and that the "anomalies" seen in the recordings could be just artifacts of the technique.[14]

LOOKING FOR AN EXCUSE

They may have used different neuroimaging technologies, but what the defense teams for Brian Dugan, John Hinckley, Grady Nelson, and Peter Braunstein all have in common is an intention to persuade a jury that their clients were not criminally responsible for their actions. Could it be that some brain injury or, more fundamentally, something about their genetics had made a crime inevitable?

As we have seen, the validity of brain imaging data remains in question, and it is notable that in several of these cases it did not win over the jury. Similar controversy has surrounded the links between genes and behavior. To what extent do our genes predispose us to act in certain ways? There seems, for example, to be an inherited component to psychopathic behavior (see box: Are You a Psychopath?). And yet not everyone labeled a "psychopath" will end up committing violent crime. The answer is clearly not that straightforward.

The field of behavioral genetics received a real boost in 1993. Careful study of a Dutch family with a history of aggressive and antisocial behavior revealed that all of the men, over several generations, carried a mutation that gave them a nonfunctional copy of the *MAOA* gene.[15] As a result of this mutation, these men did not produce a protein called monoamine oxidase A. This protein usually plays a role in degrading various neurotransmitters, chemicals that pass signals from one nerve cell to another. The absence of this "off switch" therefore meant that the affected individuals had too much of a nervous system message called serotonin, and this was said to explain their behavior.

The Dutch case is an extreme example. Other studies trying to show a link between the level of MAOA and behavior have been far less clear cut.[16] Even if a whole set of genetic factors giving someone a predisposition to aggression or impulsiveness can be identified, does this excuse them or mean that we should give them less punishment? To what extent are we "genetically predetermined"? How much "free will" do we have to override our genetic heritage?

Once again, juries have responded in rather different ways. In 2009, a court in Italy reduced a man's sentence for murder because he was found to have one of these putative "antisocial genes."[17] This happened again in September 2011, when a murderer's life sentence was reduced to 20 years after

ARE YOU A PSYCHOPATH?

There are different checklists to assess whether somebody is a psychopath. Below is the list compiled in the *Oxford Handbook of Psychiatry*. Each item is rated on a score from zero to two and the sum determines the degree of psychopathy of a person.

- Aggressive narcissism
- Glibness/superficial charm
- Grandiose sense of self-worth
- Pathological lying
- Cunning/manipulative
- Lack of remorse or guilt
- Emotionally shallow
- Callous/lack of empathy
- Failure to accept responsibility for own actions
- Socially deviant lifestyle
- Need for stimulation/proneness to boredom
- Parasitic lifestyle
- Poor behavioral control
- Promiscuous sexual behavior
- Lack of realistic, long-term goals
- Impulsiveness
- Irresponsibility
- Juvenile delinquency
- Early behavioral problems
- Revocation of conditional release
- Many short-term marital relationships
- Criminal versatility

the Italian judge considered that the neuroimaging data and the fact that she had low levels of activity of the *MAOA* gene proved that she had a mental illness.[18] Elsewhere, an opposite argument has been made; if an individual's genetic makeup means that they could never be "redeemed" and they would therefore remain a risk to society, should they actually receive a *longer* jail term?[19]

This poses another moral dilemma: does imprisoning someone on the basis of their future dangerousness cross a boundary? Does this takes us from determining criminal responsibility into the realm of predicting future criminal activity? The film *Minority Report* depicted a world in which people could be arrested by the "Precrime" department of the police for an offense that they had not yet committed. The role of the fortune-telling "precogs" in the movie would not be necessary if brain imaging or genetic analysis of an individual already revealed a propensity to criminality that puts them beyond the remedial influence of a spell in prison.

We are likely to see brain imaging and behavioral genetics playing an increasingly prominent role in the judicial system over the coming years. For the time being, however, the most robust correlation to criminal behavior is actually poor social environment during an individual's early childhood, and even here criminality is not inevitable. Modern science is providing us with clearer understanding of the genetics, the biochemistry, and the cellular physiology of the brain, but the defense that "it's not my fault, my brain made me do it" is likely to ring hollow with judges and juries for some time to come.

THE DEBATE

IN FAVOR:

- Can help communicate with patients in a vegetative state.
- Can be useful to catch liars and criminals. It can also provide an alternative to torture to extract important information from prisoners.

AGAINST:

- It may not be accurate enough to ever be used in court.
- It can be seen as an invasion of privacy.
- It can be a source of abuse.
- It can lead to ethical problems regarding the right to die of people in vegetative states.

CHAPTER 8

Playing God

A BAD DAY AT THE OFFICE

When the final report was published, the investigators concluded that a combination of poor design, inadequate training, and bad luck were the root causes of the disaster. With hindsight it was obvious that the valve for controlled discharge of the bioreactor into the sanitation plant should not have been positioned next to the sluice gate for release of uncontaminated liquid into the nearby river. But nobody had spotted this fault in the design when SynthEng BioSolutions' first plant had opened a decade earlier.

Everything had been going so well for the company, the producer of the first truly green eco-fuel, that nobody bothered revising the plans for later facilities. Why would they? SynthEng BioSolutions was a runaway success story. It was widely acknowledged to have made the single biggest contribution to the reduction in global warming. It had received countless awards and prizes for its work; it had even been heralded as "the savior of the planet."

It was, of course, easy to be wise after the event, but Michael couldn't help feeling responsible. Hell, that's because he *was* responsible! He had been the duty engineer at the Cape May plant in Delaware that night, and it was his blunder that had led directly to the release of 20,000 liters of genetically modified bacteria into the water supply. He had pressed the wrong button at the wrong time. Nothing especially serious, if you have the right countermeasures in place. But the engineers had never foreseen the possibility and so there was no means to correct Michael's mistake.

As the full scale of the disaster became apparent, the newspaper headlines had been graphic. Many drew comparisons to earlier tragedies. "Ecological Chernobyl" said one, "GM Tsunami" declared a second. Others played ironically on the name of the company: "BioSolution? BioProblem!" was one stark offering. It didn't take long for everybody in Cape May to know the identity of the person responsible for the catastrophe. Demonstrations started spontaneously in front of Michael's apartment building. Synth Eng BioSolutions had a scapegoat, for sure, but at this rate he wouldn't make it alive to any trial. So they shipped him to the farthest place they could think of: a cheap, anonymous motel in Alaska, where he was supposed to stay, safely guarded by a couple of company goons, until the waters calmed.

But they never did: the storm was just starting. Unsurprisingly, discussion in the weeks following the accident moved from the specific events surrounding the discharge of so much sensitive material to questions about the wisdom of the science itself. Defenders of the project reminded skeptics of the reduction in greenhouse gas emissions that had been seen over the decade since the biofuel factories had been established, how the cost of fuels had been reduced, and how spin-offs from the technology had developed cheaper medicines.

The idea behind the biofuel scheme had involved some truly fantastic science. The bringing together of genes from various species to create *Klebsiella synthetica*, a new man-made microbe capable of generating ethanol, remained a technological triumph. Scientists had speculated for a long time that the carbon tied up in the indigestible parts of plants would serve as an excellent alternative to fossil fuels, provided that we could work out a way to liberate this so-called **biomass**. It was SynthEng BioSolutions that had solved the conundrum, developing an organism capable of converting this waste material into a versatile biofuel.

Michael couldn't help watching the debates on TV and read the columns in the newspapers. He had managed to ruin everything good that SynthEng BioSolutions had done . . . and the company itself. Stocks started going downhill the first week after the accident, and it didn't seem like anything could stop their descent. The outlook was bleak. Michael felt bad, obviously, but also detached from the whole thing. Prisoner in his Alaskan refuge, with no friends and no comfort other than the TV and a steady supply of canned food, he was little more than a powerless spectator of the tragedy unfolding in front of his eyes. A tragedy that didn't seem to have an end.

Once released into the wild, the bacterium wreaked havoc on the Delaware Bay. By breaking down their normally indigestible materials, the

---

VOCABULARY

Bioerror: The accidental release of an engineered or synthetic organism into the wild.

Biomass: biological material, most often plants, which represent an alternative to fossil fuels as an energy source.

Bioterror: The intentional release of an engineered or synthetic organism into the wild, as part of a terrorist campaign.

Chassis: in synthetic biology jargon, a species of bacteria used to house a genome assembled in the lab.

Engineered organism: any natural organism, including plants and animals, that has been genetically modified in the lab.

Genome: the complete genetic material of an organism, contained in its DNA.

Restriction enzymes: molecular "scissors" used to cut DNA at a specified position.

Synthetic organism: an organism, probably a microbe, built from scratch in the lab.

---

synthetic bacteria created a chink in plants' armor that allowed other organisms to exploit this weakness, leading to the decimation of whole fields of valuable crops. It had spread so fast that before anyone could think of a possible solution, the artificial organism could be found from the cotton fields in the South and the orange groves in Florida to the corn fields in the Midwest and the California vineyards.

Coupled with this, the organism's ability to produce ethanol led to the pollution of water courses and the death of countless aquatic species. Now not only the farmers wanted to kill Michael: every fisherman in the eastern seaboard also had him in their crosshairs. He was actually very happy to be locked away in a remote frozen town, far from all the angry mobs.

Scientists eventually found a solution, but it took them over a year to engineer a predator that could end *Klebsiella*'s reign of terror. They called it the ultraK-T4 phage. It was a genetically modified virus from a family that only attacked bacteria. Therefore, there was no risk that it could infect humans or animals. With some modifications made in the lab that forced it to focus on only one target, the ultraK-T4 became *Klebsiella synthetica*'s worst enemy. The new microorganism was released into the wild immediately. It started diligently destroying the bacteria until, many months later, none could be found in the American waters and fields. It was fighting fire with fire, but everybody agreed that there was no other solution.

Michael witnessed everything from his secret hideout. With the collapse of the company and the secure mothballing of all its plants, it seemed that everybody had forgotten about him. Even his bodyguards disappeared overnight. Just the same, he preferred to stay in the shadows a little bit longer. The U.S. economy entered its worst phase since the Great Depression of the 1930s. The experts that Michael liked to watch debating in late night shows were pessimistic and didn't stop forecasting the end of the U.S. domination of the world and the beginning of a new era, in which the East would set the ground rules. And it was all because of a simple bacterium. Well, it was all because of him, really.

It could have been worse, he always said to himself. He could be rotting in jail, or hanging from the nearest tree. Or the bacteria could have spread to other countries, and then the crises would have had even more severe consequences. No: he shouldn't complain too much. Naturally, then he didn't yet know that once all *Klebsiella synthetica* had been eliminated from the ecosystem, the ultraK-T4 would have no prey to attack. Under this selective pressure, it wouldn't take much time for mutants to emerge. These changed viruses would be more than happy to expand their list of victims to other innocent bacteria, some of them essential to many of the processes needed to keep the fields and the waters teeming with life. Finding a solution to that was going to be much more complicated . . . .

THINK ABOUT IT . . .

Can we ever be sure that a synthetic or **engineered organism** will not escape the lab where it's being studied? What degree of risk should be acceptable? Should we stop manipulating organisms despite the numerous benefits that can be achieved if we are able to unlock their potential?

Should scientists be allowed to use new technologies if they cannot be certain of their safety? Does science need to be controlled to protect humanity? Who would have to regulate it? Where would you set the limits?

MAKING LIFE IN A TEST TUBE

In May 2010, media services around the world were buzzing with news that an American team had generated "the first man-made organism," a bacterium known as Synthia.[1] Flashy headlines such as "Biologist Creates New Life"[2] or "Artificial Life Breakthrough"[3] accompanied the announcement. The research was carried out by a group of scientists directed by Craig Venter (see box: Who Is Craig Venter?).[4] He described the microbe at the heart of the story as "the first self-replicating species we've had on the planet whose parent is a computer."[5]

Putting aside the hype that Venter's work always seems to generate, the production of Synthia really was an amazing technological feat. The scientists started by using online databases to look up the sequence of the DNA for a bacterium called *Mycoplasma mycoides*, which was chosen for having a relatively simple and short **genome**. They then assembled it from scratch.

Over the past 20 years, it has become possible to chemically manufacture stretches of DNA in the lab. The process is not foolproof: the longer the piece of DNA you want to make, the more likely that a mistake will occur. But nobody had attempted before to synthesize a *whole* genome this way. Venter's team outsourced the production to a company who then provided them with over 1,000 pieces of DNA, each about 1,000 letters long, and each of which they had painstakingly checked for errors. These were assembled in the correct order (following the information

---

WHO IS CRAIG VENTER?

Vietnam War veteran Craig Venter is one of the most colorful characters in modern science. Already successful in studying the effects of adrenalin on cells, Venter came to the notice of the public when his company Celera set itself up against the official team to sequence the complete human genome. The resulting race was formally declared a draw, with both groups publishing their results at the same time.

A brilliant media manipulator, Venter has also secured enormous amounts of money for research projects including both his work in synthetic biology and to trawl the world's seas looking for undiscovered species. To the skeptic, the latter might look suspiciously like an excuse to indulge his passion for sailing ocean-going yachts while dragging a sampling net.

THE SCIENTISTS' SENSE OF HUMOR

The inclusion of "watermarks" in Synthia's genome was useful to distinguish it from the DNA of the original bacterium. In typical Venter fashion, however, the choice of watermarks caused a few eyebrows to be raised. Within the sections of the DNA that do not contain genetic information (and therefore are not vital to life), the scientists included DNA letters that if they were being read in the conventional way would have spelled out their own names, a message in HTML script congratulating anyone decoding the sequence for doing so, and several of their favorite quotes, including: James Joyce ("To live to err, to fall, to triumph, to re-create life out of life"), Robert Oppenheimer ("See things not as they are, but as they might be"), and Richard Feynman ("What I cannot build, I cannot understand").

retrieved from the databases) to produce that complete bacterial genome. In actual fact, the DNA sequence was not quite identical to the original. Venter's team deliberately included a few changes when the DNA was being made. They refer to these as "watermarks" (see box: The Scientists' Sense of Humor).[6]

Once the DNA was ready, they needed somewhere, or more specifically some cell-like environment, into which they could put it. They took a cell from a related microbe, removed its own DNA and injected the lab-made genome into it. When it started growing and dividing, they knew they had reached the goal of creating a completely new, fully functional organism. They gave it the name of *Mycoplasma laboratorium*, but the catchier nickname Synthia was coined by Canadian environmental pressure group Erosion, Technology and Concentration (ETC) and has been popularized by others.[7] The whole process needed the efforts of 20 scientists working for over a decade, and the final price tag was $40 million.

Despite the headlines shouting about "creating new life," it is evident from what we just said that we can't really call Synthia "new." It was a forgery, a mix of one existing bacterium with the genetic information of another. But what's the real use of all this? Venter's experiment was all about "proof of principle"—to look at the feasibility of this approach for future times, when the assembled DNA would have been more radically altered to suit our needs. This is potentially a key step in the production of purpose-built microbes.

AND NOW, DESIGNER BUGS

There are certain reactions that are difficult to achieve in the chemistry lab but that bacteria can carry out with ease. It makes sense to take advantage of this. Putting bacteria to work for us is not such a radical idea as it may seem. In fact, mankind has exploited the capabilities of microorganisms for centuries without necessarily knowing the scientific principles involved. Microbes are used in baking and brewing and in the manufacture of cheeses and yogurt. More recently, we have exploited them in the treatment of sewage and in the production of antibiotics. As Venter's experiment indicates, however, we may be moving into an era where we go from taking advantage of the natural properties of microorganisms to creating bacteria designed to do particular jobs for us. The potential applications of this approach are enormous. Possible uses include the manufacture of pharmaceutical drugs, the clean-up of pollutants, and, as in our scenario, the production of biofuels.[8]

To create a new organism you can start from scratch with just a cocktail of chemicals or use bits and pieces already found in nature. The first approach is obviously more difficult—to bypass the work of millions of years of evolution and reimagine life. Unsurprisingly, the few teams taking this ambitious approach have so far had little success in the laboratory.

The alternative looks more feasible. As with the work conducted by Venter's group, you first need to design the genome and then put it into a suitable cell (what is usually called a **chassis**).[9] Some researchers are looking for the ideal organism to be used as a universal chassis (one that could be used in any experiment), while others are starting to establish the basic combinations of genes that can keep the microbes running and doing their job. For instance, scientists from the Massachusetts Institute of Technology have pioneered the "BioBricks" scheme,[10] establishing a catalogue of preconstructed gene combinations for specific purposes that interested researchers can assemble into biological projects in much the same way that an electrical engineer might select a particular component to incorporate into a circuit.

To serve as a chassis, a cell needs to be able to live and replicate on its own. The kind best suited for this job is a bacterium. Viruses tend to have smaller genomes, which is why a virus would appear superficially to be a better starting point (the first synthetic virus was, in fact, generated in August 2002).[11] However, they are not self-sufficient because they are dependent on the need to hijack the machinery of any cell they invade in order to complete their replication. Bacteria are therefore a better solution.

In principle, scientists could use any bacterium as the basis for their bioengineering, but different species can vary widely in both the number of

MINE SWEEPERS

Landmines are a scourge of modern warfare. In many countries around the world, hidden mines can cause devastation for many years after a conflict has ended. Scientists at Edinburgh University have used the BioBrick approach to develop bacteria that glow when they encounter explosives leaking from a buried weapon. The bacteria could be sprayed onto a suspected minefield from the air, and once the land-mines have given away their hiding place, they would be much easier to remove.

genes that they have and the ease with which they can be grown in the lab. The ideal, therefore, would be a quick-growing cell with a minimum number of genes necessary to survive. Researchers could then add specially se-lected combinations of genes in order to produce a cell capable of fulfilling a particular function (see boxes: Mine Sweepers and Fighting Malaria with the Help of Synthetic Biology).[12]

In the case of Synthia, they initially chose a bacterium called *Mycoplasma genitalium* as their chassis. It has a remarkably small genome, consisting of only 480 genes (for comparison, *Escherichia coli*, the most common bacteria

FIGHTING MALARIA WITH THE HELP OF SYNTHETIC BIOLOGY

Malaria has killed more humans than any other disease. It still infects some 500 million people each year, resulting in about 2 million deaths. The battle against malaria has been hampered by the spread of resistance to traditional antimalarial drugs. An exciting new medicine, artemisinin, was found in the plant *Artemisia*. Despite being highly effective against malaria, the need to extract and purify this compound from plants was a significant limitation on its use because the yield was so low. In 2003, scientists in Berkeley, California, successfully transferred the genes needed to make a precursor form of artemisinin into a bacterium. Although some further processing is still needed to get the active drug, this step allows for industrial scale-up, making artemisinin available to many more people.

used in research, has around 4,300 genes). Over several years, Venter and his team worked systematically to see how many of these genes were really essential to sustain life. They were able to reduce the list down to a remarkably small 380.[13] The problem was that *Mycoplasma genitalium* naturally grows very slowly. It takes between 12 and 16 hours for one cell to divide and become two cells (*Escherichia coli* takes just 20 minutes). Therefore, despite the attraction of the ultrasmall genome in *Mycoplasma genitalium*, they shifted their attention to using *Mycoplasma mycoides*, which has more DNA but grows substantially faster (it doubles in about 80 minutes).

The rapid pace with which genomes have been sequenced and the success in creating Synthia have engendered a real excitement that we stand at a pivotal point in being able to tailor bacteria to our needs. For the most part, however, the fulfilment of these dreams is still likely to be several years away.

A GREAT IDEA . . . OR A RECIPE FOR A CATASTROPHE?

In our initial scenario, a bacteria species designed by man was fulfilling a valuable role in converting waste biomass into a versatile fuel.[14] All was great, while it was kept within a carefully controlled, industrial setting. However, everything changed when it was inadvertently released into the environment. The fallout depicted in our story is somewhat exaggerated (we hope), but a purpose-built microbe could indeed cause a disaster in the wrong setting. Are they really that dangerous?

BEGINNERS' GUIDE TO GENETIC ENGINEERING

The ability to manipulate DNA relies on the use of so-called **restriction enzymes**. These proteins are normally found in bacteria, where they are probably used as a defense mechanism to chop up the DNA of rival microbes. Restriction enzymes recognize a specific string of six or eight letters in the genome and then cut the DNA. In the early 1970s, scientists started to realize that under the right conditions, they could use these enzymes to cut two different pieces of DNA and then connect them. This allowed them to "stitch" together fragments of DNA in the desired order. Restriction enzymes from different species of bacteria recognize different strings of DNA letters. Most modern biology laboratories will have whole libraries of different enzymes, which allow scientists to cut DNA at exactly the position they want.

JUST IN CASE . . .

Aware of the potential problems, scientists are taking steps to offer some protection against the impact of accidentally released microbes. One safeguard is to make the engineered cells dependent on the controlled supply of a particular nutrient that is relatively rare in the natural environment. In this manner it should be possible to ensure that any bacteria escaping from a laboratory will not be able to survive and thrive in the wild.

The development of **synthetic organisms** shares some interesting characteristics with scientific advances in quite different fields. Take, for example, the application of atomic energy. Nuclear power stations can serve a fantastic role in fulfilling the energy needs of a large city, but when things go wrong, it can be an environmental disaster of a frightening scale. Incidents like Chernobyl in 1986 and Fukushima in 2011 are dreadful reminders we may never be able to control all the variables we need to make sure that accidents don't happen. Is the use of nuclear energy worth the risk? This is a hotly debated question at the moment, which may serve to inform discussions about synthetic organisms. The safety of science is, in fact, a more generic question. To what extent should scientists be aware of all the potential risks of an experiment? No research is entirely risk-free, but at what point do the risks outweigh the benefits? Where should we set the limits?

In the case of the release of new life forms into the wild, there are a few options to consider. One is the deliberate release of harmful organisms by a malevolent group or individual, called **bioterror**, which we will discuss in the next section. The second is **bioerror**.[15] This may be the result of an accidental release, but it might also result from unanticipated consequences of a planned introduction of organisms into new environments in order to achieve a specific goal, as we have explained in the second half of our story. The history of science is littered with examples when this has not gone quite according to plan. The following cases are illustrative of the problem.

In the 19th century, horticulturalists in Europe and America took a shine to Japanese knotweed, a plant they had found in Asia, and added it into their ornamental gardens. It was also valued as a source of cattle feed, and honey producers appreciated the fact that it flowered at different times

of year than native plants. Unfortunately Japanese knotweed quickly became an environmental menace. It grows up to 10 cm a day and has a root
system that is both formidably difficult to remove and also strong enough
to destabilize buildings. Governments have been spending substantial time
and money since then to keep knotweed at bay.[16]

In the 1930s, Australia was looking for a way to suppress the population
of beetles ruining their fields of sugar cane. Someone hit on the plan to
import cane toads from Hawaii. Unfortunately, the toads demonstrated a
preference for feeding on birds' eggs, rival frogs, and a range of other in-
sects rather than the intended beetles. They also proved to be fatally poi-
sonous when eaten by various native predators, including crocodiles. Not
an overwhelmingly successful plan.

Some introduction of non-native species has not been so deliberate.
Mink, a carnivorous mammal originally found on the American continent,
has been imported into Europe and farmed for the fur industry. Over the
years there have been frequent stories of accidental escapes or releases by
well-meaning, but badly informed, animal rights activists. In October 2009,
coordinated attacks on four fur farms in Northern Spain are reported to
have liberated more than 17,000 mink during one night.[17] Similar incidents
have occurred in other European countries, including France, Greece, and
the Republic of Ireland. Mink can thrive in the wild, where they feed on
small birds, eggs, fish, and small mammals such as voles and shrews. As a
consequence they are a substantial threat to the ongoing survival of native
species. They have, for example, been responsible for the significant decline
in the numbers of terns and gulls in Scotland.

THE FEAR OF BIOTERROR

Synthetic biology is a classic example of a so-called dual-use technology,[18] one that could be used both for good and for harm. Are scientists in
their quest to produce novel microbes inadvertently providing new weapons to would-be terrorists? How can we be protected from the deliberate
misuse of this technology? One solution might be to conduct key experiments in secure facilities, away from prying eyes. This has been the model
adopted to keep terrorists from gaining insights into nuclear physics, another discipline where the risk of harm is evident.

For the majority of people working in science, however, there is an expectation that you will not only tell other people what you have done but
also share sufficient details about how you have done it, so that they will be
able to repeat the experiment.[19] This is an integral part of the ability of science to progress as fast as it currently does: a secretive approach would

MAIL-ORDER MAYHEM

Over the past decade, a number of commercial enterprises offering to make specified DNA sequences for scientists have sprung up. As we have said, it was precisely this kind of service that Craig Venter and his colleagues used to order the DNA units that they used to construct Synthia. You might hope that the companies making DNA would routinely perform checks to ensure the credentials of people ordering their materials. However, a series of tests run by scientists and journalists have shown that this is not the case.

In 2006, James Randerson, of the *Guardian* newspaper in London, was able to order a section of smallpox DNA and had it delivered to a residential address. Smallpox is a virus that has killed up to half a billion people but has already been eradicated; a resurrected smallpox could be very dangerous in the wrong hands. The company making the DNA did telephone to check why delivery was being made to a private address, but when the journalist explained that his fictitious company was in the process of moving from one commercial premise to another, they were perfectly happy to send the material out to the address provided. Also, they had done no checks to investigate what the sequence might actually be and whether its use could be dangerous.

seriously slow down the rate of new discoveries. The result is a clear threat: anyone could have access to potentially destructive information. Bioterrorists might even be individuals working in garage-scale laboratories, so-called biopunks and biohackers, rather than formal scientists working in state-sponsored programs (see box: Mail-Order Mayhem).[20] The deliberate sharing of knowledge about synthetic biology could pave the way for exploitation of these methods to manufacture and release novel killer microbes against which we would have no defense.

Some have argued that this is not such a substantial threat as it seems. According to them, the dangers of terrorists turning to synthetic biology as their means of developing an attack have been overstated. A strategy based on emerging biotechnology is more difficult to achieve and not as cost-effective as, for instance, the culturing of aggressive microorganisms that might easily be harvested from the wild. Therefore, it looks like the main potential danger of synthetic biology is the accidental release of the new microbes, just as we described in our story. The pace of innovation in the field is likely to be determined by the balance between what can be achieved with these organisms and how tight are the security measures that we can put in place to protect ourselves.

GOING TOO FAR?

Scientists have been manipulating life in the lab for decades. Nowadays it is relatively straightforward for researchers to add or remove specific genes from organisms. We tweak viruses and bacteria in all sorts of ways to speed up our discoveries. We can engineer bacteria so that they produce human insulin, to the daily benefit of thousands of diabetics. We can introduce genes from different species into animals such as mice in order to study their function. We have genetically modified crops that are being grown all over the world and end up in the food chain. And as we have seen in previous chapters, we can make perfect genetic copies of a surprisingly large number of organisms. Why would creating a life form from scratch be any different? Is it more useful than modifying an existing one? Is it really more dangerous? Certainly, the accidental release of a completely new microbe should not be more alarming than a modified virus escaping a lab where it's being studied. Then why do we fear synthetic biology so much?

The media coverage accompanying the production of Synthia triggered significant debate about the ethics of this blossoming scientific discipline. It even led Barack Obama to initiate the first gathering of his Presidential

---

ASILOMAR, 1975

When genetic engineering began in the mid-1970s, the main concerns about safety were raised by the scientists themselves. They put in place a voluntary moratorium on conducting experiments using the new technology (known then as recombinant DNA). In February 1975, over 100 leading experts met at Asilomar, California, to pause and reflect on the potential consequences of embarking on this work. Their discussions included the dangers of accidental release of modified organisms, and they put in place a classification system for potential risks (and the necessary degree of associated precautions). This system has worked extremely efficiently, and some observers have called for an Asilomar-style meeting specifically to discuss synthetic biology.

Commission for the Study of Bioethical Issues.[21] When in December 2010 they reported on their deliberations, the commission decided that the developments in synthetic biology do not currently raise any novel ethical questions over and above the extensive debates that have taken place since the emergence of genetic engineering in the 1970s (see box: Asilomar, 1975).[22] They suggested that the best policy to follow at present is one of "prudent vigilance."[23] In doing so they attempted to walk a middle ground between those who would argue that we should not carry out work of this kind until we are certain that it is safe and those who feel that the benefit of the doubt ought to be given to scientists until we have good reason to think otherwise. The commission is in agreement with most experts, who have decided that concerns about potential for both bioerror and bioterror associated with genetically modified organisms have already been given thorough evaluation over the past 40 years, and that synthetic biology should not be given special consideration. But on March 2012, 111 environmental organizations considered that this report was not objective enough and that a moratorium was still needed.[24] The debate is not yet closed.

IT'S LIFE, BUT NOT AS WE KNOW IT

For some observers, the development of synthetic organisms raises the ultimate philosophical question: *What is life?* If something is assembled by man from raw materials, can it be called "alive"? Even if it looks and functions just like a similar natural organism, the fact that it doesn't come from another being but from a series of chemical reactions, they might contend, puts it into a special category. After all, this would contradict an old

---

THE ORIGIN OF LIFE

Atheists have taken the Synthia experiment as the proof that life can exist without the need of a "spirit" or a "vital spark" that starts it, a concept that is common in many religions. In their view, human beings could one day be generated from a fully synthetic egg fertilized by a full synthetic sperm, which would show that we are not much more than a (very complex) biochemical machine. Synthetic biology indicates, after all, that what we call life can indeed be generated from inanimate chemical components.

scientific principle, as enunciated by William Harvey in the 17th century: all life arises from existing life. Are we actually redefining life itself? Are we taking science one step too far just for the sake of proving that we are able to do so? Questions about the property of man-made species have also received careful consideration. Can a scientist really claim "ownership" of a living creature? Is it even moral to create a new being just to suit our needs?

Some scientists consider that creating a synthetic organism is justified simply on the grounds that it provides an excellent vehicle for fundamental research into how life really works and to identify the elements that are truly essential for life to happen.[25] In their eyes, the quest for knowledge is reason enough. Others argue that this is an extreme example of scientists "playing God" and straying into territory where they have no right to venture.[26] So far, the real-life benefits that synthetic biology promises are not clearly superior to those that can be obtained by manipulating existing organisms. Should we concentrate in other areas equally promising and less controversial? Synthetic biology is a discipline that is just starting to deliver its first important results. It's too early to see how productive and enlightening it can be, and whether its impact will balance the moral concerns that it is generating. As we have seen, we are still far away from being able to really create life from scratch. This should allow us plenty of time to discuss these matters more carefully.

THE DEBATE

IN FAVOR:

- Possibility of creating new life forms that could be useful (such as cleaning oil spills, generating energy, production of pharmaceutical drugs and other products).
- Lessons learned with these technologies could be applied to other areas of research in genetics, microbiology, and so forth.

AGAINST:

- Danger of creating new organisms that escape from our control and unsettle the natural ecosystems.
- Danger of rogue scientists creating harmful organisms on purpose to be used in terrorist attacks.
- We may not gain more than just manipulating existing organisms.

CHAPTER 9

Trust Me, I'm a Scientist!

In our discussions so far, we have reflected on the science and ethics of various specific developments in biology and medicine. In this final section we are going to take a step back to think about some of the ethical issues associated with actually "doing" science.

JUST THIS ONCE

As Phil collected his samples from the fridge, he was aware that it was the third time this week that he'd be working late at the lab. He'd wait a few minutes before ringing Claire to let her know. He needed the time to steel himself for her response. It wasn't her disappointment that he found so difficult to face—it was her weary resignation that once again he'd abandoned her to sort out the kids' bedtime on her own. He did feel guilty about leaving her in the lurch once more, of course, but what else was he to do?

It was now six months since he had been called in to see the director of the institute, Professor Sanderson. It had been advertised to Phil as a routine performance review, but the meeting had been altogether more menacing than he had foreseen. Never a man of many words, Sanderson cut straight to the chase. He reflected that the Crombie Institute had a worldwide reputation for the quality of the science they conducted. He reminded Phil that, as befits a facility of their stature, they had provided him with an extremely generous start-up package to get his lab off the ground. And Sanderson pointed out that in the two and a half years since Phil had joined the institute, he had not published a single research paper. This state of affairs would not be tolerated for much longer; Phil had one more year to get a paper accepted into a major scientific journal or, with regret, they would have to let him go.

Now, Phil sat in the lab, a sense of deep foreboding washing over him. It wasn't just the humiliation of knowing that his professional career was about to come to a juddering halt. If he lost this job, his family would have to move out of their current apartment for sure, probably away from New York altogether, to wherever he could find another position. And it wouldn't be easy to secure a new role anywhere after being expelled from the Crombie Institute. He knew how much Claire loved living on the Upper East Side, and the kids had settled so well into their kindergarten . . . they'd all made such good friends in the neighborhood! Having to move into a small, remote city would be a serious blow for the whole family. He couldn't disappoint them.

To be honest, his research over the last few weeks had gone pretty well. A lot of the data his group had gathered fitted his hypothesis, but Phil was still struggling with the final experiment needed to secure publication in a top journal. He'd tried it three or four times in recent days, but the results had always been inconclusive.

When he'd decided to stay on this evening, Phil had fully intended to repeat the elusive experiment one more time. However, the urgency of his predicament had suddenly become all too clear. All scientific research papers need to be read by a panel of other scientists before they can be published. Factoring in the time it would take for his manuscript to be evaluated by these reviewers, which is usually months, he calculated that he now had only a few days to get the necessary result if he wanted to make Sanderson's deadline.

In that instant, Phil was struck by an alternative solution—he would create the data using Photoshop. Why not? Phil wouldn't have described himself as an expert "Photoshopper," but he'd used the software often enough to have a pretty decent idea of what he needed to do. He could take an image from one of the unsuccessful experiments and simply add the missing information. He'd never done anything like this before, of course, but desperate times called for desperate measures.

Phil's heart was pounding and his brain went into overdrive. He was already formulating arguments to justify the course of action he had now

VOCABULARY

Fabrication: creating data that does not exist.

Falsification: cherry-picking the data that you show in order to make it fit in your hypothesis.

Grant proposal: a document that explains what the scientist plans to do with the money if they receive it.

Peer review: process of scrutiny to which most scientific papers are subjected before being made public. An editor picks two to four experts in the field, who will remain anonymous to the authors. These reviewers comment on the manuscript and decide whether it can be published or the data needs to be strengthened first. The editor has the final word.

Placebo: a mock drug without any of the active ingredient being analyzed.

Placebo effect: the psychological effect of being involved in an experiment, even though you haven't been given the drug being studied.

Plagiarism: copying somebody else's results and claiming authorship.

resolved to take. It was not, he told himself, as though he didn't have any data—his results already told a consistent and plausible story. It was just that single experiment that refused to show what it was supposed to. Somewhere there was a technical glitch that he couldn't see and, who knows, it could take him months to spot it. He just couldn't afford that. What he was doing now was simply speeding up the process of getting the final image for the paper. It wasn't exactly cheating. He was 100 percent certain that the photo he was about to generate was exactly what the experiment would show if only he had time to perfect it. And while reviewers were going over the manuscript, he would have plenty of time to get it done for real, to confirm that he was indeed right. This was just skipping a few steps out of sheer necessity, not actually fabricating data. There was nothing really wrong with it.

Satisfied with that self-justification, he stood up to put the samples back in the fridge. But as he did so, a further realization hit him. Everyone in the lab knew he'd been trying desperately to get the procedure to work. If he told them in the morning that he'd finally cracked it, then the ongoing presence of the sample tubes might arouse suspicion. He had to cover his tracks. Instead of putting them in the fridge, he slipped the tubes into his coat pocket—he'd drop them off in a trashcan on the way home. No big deal.

Phil moved instead to his computer, opened a data folder, and chose a photo showing the results of an experiment he had carried out at the end of the previous week. The image showed a series of black bands on a white background. He just needed to add one more and the last piece of the puzzle would fall right into place.

Phil picked a band from a second file, one that could credibly substitute for his missing data. He selected the region of the photo that he wanted and copied it. He returned to the original photo, pasted in the important band, and used the direction keys to move it into the right place. It felt at home in its new position

He was shocked by how plausible the data already looked. The sharp, straight edges around the imported section would give away his deception, but the overall impression was better than he might have imagined. To deal with this issue, Phil tweaked the contrast and brightness of the added material so that it matched the rest of the picture. He then used the paint brush tool in its finest setting, working carefully along the edges with a circular motion and blending the two images seamlessly into one. It took him a while to get it right.

Phil leaned back on his chair to look at his handiwork. The result was staggering. The photo gave no hint of the trickery that had been used in its creation. No reviewer should be able to spot it! He saved the image file and turned his mind to the practicalities of completing the deception. To ensure that the truth didn't seep out, Phil would have to develop a consistent story. He wouldn't tell anyone, not even Claire, about what he had done. He didn't plan to make a habit of this kind of thing; it was—he told himself—a one-off, the lesser of two evils. If the paper describing the research was accepted, and he was sure that it was going to happen, then it would see him safely into the next phase of his contract at the institute. His future, and the future of his family, would be fine.

> **THINK ABOUT IT . . .**
>
> Given that his career and the safety of his family were at risk, was Phil's action justified?
>
> Can the pressure from the director be considered unfair, given how close Phil was to completing his experiments?
>
> Can this be considered cheating? Phil was not making data from scratch, just adding some details he was certain that he would be able to reproduce later . . . But what if he was wrong?

IT'S TOUGH, BEING A SCIENTIST

Our lives are surrounded by the fruits of scientific endeavor—the gadgets we take for granted, the medicines we need when we are ill, even the very fabrics that we are wearing owe much to the work of scientists. Modern science is a multi-billion-dollar global enterprise employing scores of researchers. Before thinking about situations where scientists may be guilty of unethical behavior, it is important to understand how "doing science" usually works.

Up until the last century or so, science tended to be somebody's hobby rather than their job. For instance, Gregor Mendel, who set the basic laws of genetics in the 19th century, was a monk who used his spare time to study how the features of plants were inherited. And Charles Darwin could spend his days working on the theory of evolution without having to worry about generating an income, since he had been born in a wealthy family. Enthusiastic amateurs made observations about the world, and if they shared these with other people at all they tended to do so in books or papers written by them as individuals. Nowadays scientists are much more likely to work in larger groups, sometimes spread over different countries and continents, with members bringing expertise in some particular technique or method to complement the skills of others within the team. Almost without exception, participation in science will be their exclusive job and they will require payment for their efforts. Somebody wanting to carry out research, therefore, needs a sizeable budget to bankroll the enterprise, covering salaries and in some cases the very expensive reagents or equipment needed for the experiments. Nowadays, funding tends to come either from a company, from government money, or from charitable donations.

In the case of government and charitable funding, scientists usually have to apply for research money by writing a **grant proposal**. This document explains what the scientist plans to do with the money if they are given it. As well as an outline of how the money will be spent, the organizations offering funding will probably want some preliminary results to show that the project is not nonsense. The application process is very competitive, and all researchers would expect to have a sizeable percentage of their grant applications turned down.

The importance of telling other people about your experiments should not be overlooked. It's actually the only way you have to demonstrate that you have kept the promises and predictions made in your grant application. And you will most definitely need that if you are planning to ask for more money in the future. Not for nothing is there a famous saying in science: "Publish or perish." As Phil's situation in our story demonstrates, there is a significant threat that without enough evidence to show that you've made worthwhile use of the money you've received, your research career might come to an abrupt end.

When it comes to the point that scientists think that they have enough evidence to share with the world, they have to persuade a small group of fellow scientists that they agree. This is the process known as **peer review**. When you submit a manuscript to a scientific magazine, one of the editorial team will send it on to two or three other scientists with relevant expertise. There are several possible outcomes to this review process. Some research will be accepted straightaway for publication. More likely, a paper will be accepted with minor changes. The reviewers might decide that there is the kernel of an interesting and relevant story here, but bounce the paper

A DOUBLE-EDGED SWORD

The peer review process itself can also be tainted with misconduct by the reviewers. Reviewers are likely to be scientists working in a closely related area of research. Therefore, there is every possibility that the reviewer will know the author of the manuscript personally and will count them either as a friend or as a rival. In the former case, they may look too favorably on the work. The bigger problems come with the latter, however, and the temptation for the reviewer to steal ideas from their rival or to make out that the work is much worse than it really is to stop it getting published. Unfortunately, you wouldn't have to talk to many scientists before you uncovered stories of being on the wrong end of this kind of experience.

back to the authors until they have carried out one or two additional experiments. Finally, if the reviewers decide that the paper is so badly written, that the data is so full of holes, or that the subject matter of the work does not map onto the interest areas of that particular journal, then they may reject the paper outright. Done properly, peer review is an important deterrent against scientific misconduct; but done badly it can be part of the problem (see box: A Double-Edged Sword).[1]

SCIENTISTS BEHAVING BADLY

It would be nice to think that all scientific research was carried out to the highest possible standards at all times and that all published data was either correct or, at worst, that any mistakes had come about accidentally. Science, however, is a human activity, and, as we are all too aware, people in all walks of life can give in to temptations to act inappropriately. Scientists are no different.

Over the past few years, several high-profile cases have shone a spotlight onto some murky goings on in science. There have been particular concerns about three potential problems: fabrication, falsification, and plagiarism.[2] **Fabrication** involves the making up of results (see box: The Magic Marker).[3] The story of Hwang Woo-Suk (see box: Hwang the Fraudster, Chapter 3) is a good example. Rather like Phil in our scenario, some of the photographs describing Hwang's results had been edited and some images even reused. In one British case, it was found that two different papers published in the *British Journal of Obstetrics and Gynaecology* by Malcolm Pearce involved imaginary patients. Suspicion was raised about one study reporting trials of a hormone-based treatment for women

---

THE MAGIC MARKER

The public has become more aware of research misconduct in recent years, but it is not a new phenomenon. In the 1970s, William Summerlin was working at the prestigious Sloan-Kettering Institute in New York. His research into ways to overcome tissue rejection included the apparently successful transplanting of fur from black mice onto white mice. The effect was rather spoiled when it was discovered that the black fur could be washed off; Summerlin had simply used a black marker pen to color in patches of fur on the white mice.

experiencing repeated miscarriages because he included data from more sufferers of polycystic ovary syndrome, the focus of the project, than he could possibly have recruited. In the second, he described groundbreaking work in which a woman with an ectopic pregnancy had the embryo successfully transferred to her uterus. It later transpired that this patient never existed.[4]

There is also fabrication in the story of Scott Reuben, a former professor of anesthesiology at Baystate Medical Center in Massachusetts. This time the consequences were worse. In June 2010 Reuben was sentenced to six months in prison when he admitted that he had made up the results of clinical trials in the area of pain management.[5] Following his confession, 21 fraudulent scientific articles were retracted. Another scientist who ended up behind bars was Eric Poehlman, who worked at several institutions including the University of Vermont, then was jailed for a year in 2006 when he admitted making up data in papers and in grant applications relating to aging, menopause, and hormone-replacement therapy. As well as going to prison, Poehlman had to pay back about $200,000 (although this is a mere fraction of the $2.9 million he received as a result of his questionable grant applications).[6]

Researchers who succumb to fabrication are frequently guilty of a second category of research misconduct: **falsification**. In the case of falsification, the scientist has some genuine data but is selective in his reporting of the results. Results that fit with their story are included, whereas some other data that doesn't support the author's case are left out. This is sometimes hard to spot without access to all of the original data, and it is one of the main reasons that there are increasing calls for codes of conduct regarding the preparation and storage of laboratory notebooks, in which every result from every experiment is detailed.

Finally, **plagiarism** differs from both fabrication and falsification in that it describes a situation in which there are indeed some genuine scientific results—the only problem is that the work does not belong to the person trying to claim the credit (see box: How to Deal with a Cheat).[7] One recent high-profile case of plagiarism involved the British psychiatrist Dr. Raj Persaud, who was accused of stealing large sections of his 2003 book *From the Edge of the Couch* and various other scientific publications. And an article of his about the famous psychologist Stanley Milgram was retracted because it was very similar to a work by American academic Thomas Blass. At a hearing of the General Medical Council, the group of medical professionals that polices their profession in the United Kingdom, Persaud eventually admitted to a failure to adequately acknowledge his sources and was suspended from work for three months.[8]

HOW TO DEAL WITH A CHEAT

Regulations and attitudes regarding scientific misconduct vary in different countries. In March 2011, Karl-Theodor zu Guttenberg, the defense minister of Germany, was pressured to resign after it was found that he had plagiarized large sections of his PhD thesis.

Around the same time, it was discovered that Joana Ortega, vice-president of the Catalan autonomous government in Spain, had falsified her curriculum vitae to show she had a degree in psychology, although she had not in fact completed the courses. Despite the public uproar, Ms. Ortega argued that it had just been a mistake, and she still holds her position today.

Also in March 2011, it was found that Mariastella Gelmini, the Italian minister of education, had cheated to pass her bar exams. The academic community complained that this was setting an extreme bad example, but she didn't resign either.

In all these situations it was debated whether having a degree (or having lied about it) had any impact on how well they were performing in their jobs and whether they deserved to be reprimanded.

YOU'LL NEVER GET AWAY WITH IT, DR. JONES!

What can be done to reduce the likelihood of research misconduct? Several strategies are being put into place. First, there is a greater emphasis in the training of new scientists to instill good practices. Second, research institutions are being encouraged to undertake periodic audits of their scientists to ensure that there is no scientific or economic fraud taking place.

When it comes to checking the validity of submitted manuscripts, more scientific journals are now routinely using sophisticated tools to unearth potential problems. Where previously journals required whistleblowers or particularly sharp-eyed reviewers to root out foul play, technological solutions are now at hand. Software such as CrossCheck and DEJAVU can show whether the research being described has actually been published before, either by different scientists or by the same scientists trying to boost their résumés by writing more than one account of the same experiments published in different journals to inflate their apparent productivity.

Statistical tests can be used to see if the data in a manuscript is genuine, because there are measurable differences between real randomness and numbers chosen by someone trying to be random. Many journals are now also carefully checking the images in submitted papers. The Hwang Woo-Suk cloning case and other high profile examples have seen magazines go from a default position of trusting the authors unless something is clearly suspicious to a much more proactive scrutinization.[9] In the initial scenario, Phil could have gotten away with the Photoshopping of data images if he had tried it 10 or 15 years ago, but probably not now. He may not have been able to spot telltale signs of his creative artwork by eye, but his attempted fraud was crass and amateur, and he would have been eventually caught.

Naoki Mori, a prominent Japanese virologist, is one person to have recently been caught trying a stunt of this kind. Mori was dismissed from the University of the Ryukyus, Japan, in January 2011 after being found guilty of manipulating images in over 10 papers. These were retracted from the scientific journals and Mori was banned from publishing in any journal produced by the American Society for Microbiology for a further 10 years.[10] When interviewed, Mori insisted that the manipulated data were just control experiments and that his actual data still stand, which sounds similar to the justifications that Phil was adopting in our story.

Despite these examples of misconduct, the majority of scientific research is carried out to high standards of integrity. Most scientists are motivated by genuine curiosity about the world around us and work hard to design the right experiments and conduct them in an appropriate way to obtain data that is as accurate as possible. Of course, genuine mistakes will sometimes be made. According to a recent study, in fact, less than half of all retractions are due to real misconduct. The rest are due to innocent errors introduced by scientists overly hasty to share the fruits of their labors. It's also interesting to note that only 4 percent of all retracted papers had sponsorship from a pharmaceutical company, and yet "Big Pharma" money is still seen as one of the main drivers leading scientists into misconduct.[11]

In summary, in situations where someone may have been tempted to try to cheat, there are now increasingly sophisticated checks and balances that can identify deception. Also, the penalties for getting caught are severe: loss of employment, suspicion regarding your previous work, some of which may be officially discredited and disowned by the magazines where it was published, and, as we have seen, substantial fines and even loss of liberty.

MUCH MORE THAN A REPUTATION IS AT STAKE

The impact of scientific misconduct affects more people than simply the professionals involved in it. Due to the nature of biomedical research, it can also have serious consequences for public health. One example is the case of Andrew Wakefield. In 1998 Wakefield and colleagues published a study apparently linking the triple MMR vaccine (against measles, mumps, and rubella) and autism, a neurodevelopment disorder in which the person struggles with social interaction.[12] The paper had a huge impact in the media and was grist for the mill for campaigners arguing against the immunization of children. The involvement of several celebrities fueled the publicity regarding Wakefield's work.

Right from the outset, various specialists were skeptical about the alleged connection between MMR and autism. Over time more and more evidence refuting the link was generated in the medical literature, but the general press kept peddling the story. Eventually some newspapers and documentaries started to pick up the weaknesses in Wakefield's case. When they did so, it was not just the lack of reproducibility of his results that were of concern. His methods for recruitment of children for the study, his vested financial interests in finding a connection, and evidence that some of the data were selectively reported, if not outright faked, all cast doubt on the MMR and autism story.[13] In 2010 Wakefield's original paper was retracted by the journal that had published it, and in May of the same year he was struck off the register by the General Medical Council for serious professional misconduct.

Unfortunately the issues are rather more serious than the gradual unwinding of Andrew Wakefield's career (which actually continues via various clinics in America). The real problem stems from the period of 12 years or so during which concerned parents stopped giving their children the MMR vaccine. The percentage of vaccinated children in the United Kingdom went from close to 98 percent in the 1990s to 78 percent, with some areas like London falling to 50 percent, well below the levels necessary to maintain so-called herd immunity (if there are enough children vaccinated, a disease won't be able to be transmitted effectively, and even those not vaccinated will be effectively protected). In consequence there has recently been a spectacular increase in cases of infectious diseases that had been close to eradication. For instance, in the four years leading up to the publication of Wakefield's contentious paper, there had been respectively 27, 48, 100, and 67 confirmed cases of mumps in England and Wales. Since 2003 there have never been fewer than 1,000 cases a year, and in one year 6,021 cases of mumps were confirmed.[14] These are not trivial infections; measles, mumps, and rubella can be fatal. Wakefield's misconduct has almost certainly led directly to unnecessary deaths. This shows innocent people can suffer when scientists decide to cheat, and underscores the importance of increased regulation of research.

SECRECY AND ABUSE

Disastrous as Wakefield's vaccine fiasco was, there have unfortunately been cases of scientific misconduct in history with far worse impact in humans. There is a litany of examples where fundamental standards for research have not been maintained. History is rcplctc with stories of so-called scientists performing experiments on people perceived as "inferior," such as prisoners, slaves, or handicapped people (see box: War Criminals). Nowadays it would be practically impossible to see situations as extreme as that again, but the global reputation of scientists has been permanently tarnished by such horror stories. The fear of government-sponsored secret experiments is still present in the collective psyche. Suspicions also persist that some pharmaceutical companies may be guilty of conducting unethical trials in developing countries (as dramatized in the book and film *The Constant Gardener*), despite little available proof.

One of the most infamous examples of the exploitation of research subjects was the Tuskegee syphilis experiment.[15] For 40 years from 1932, the U.S. government studied the effects of syphilis in black men in Macon

WAR CRIMINALS

Human experimentation has been common in past wars. Japan's infamous Unit 731 used Chinese prisoners to test biological weapons during World War II. Nazi experiments on inmates of concentration camps have also been detailed in many books and reports. Most of these atrocities were performed not only with a frank disregard for human rights but also with dubious scientific reasoning. Thus, very few useful conclusions ever came out of the work.

The case of Pernkopf's *Atlas of Anatomy* poses a particular ethical dilemma. Edward Pernkopf, from the University of Vienna, started working on his *Atlas* in 1933. His illustrations, taken from the careful study of human corpses, are among the best regarded in the field. The problems arose in 1998 when it emerged that Pernkopf's department had been receiving bodies from prisoners executed by the Gestapo. Should we disregard one of the best anatomy books because of the unethical way the subjects were obtained? Or should we value the book for its evident scientific contents, independently of the crimes that made it possible?

County, Alabama. The subjects were knowingly left untreated to allow for study of how the disease progressed, despite the fact that, from 1947 onwards, doctors knew penicillin could have cured them. The victims, who were from a poor social background, were never informed of the reason of the study or even that they were sick. They were even prevented from receiving treatment when the disease was diagnosed in routine medical exams. As a consequence, many men died of syphilis, many wives were infected, and many children were born with the congenital form of the disease. In 1972, a leak to the press put a stop to the experiment.

The Tuskegee experiment is one of the reasons why regulations regarding the participation of human subjects in clinical trials and experiments have been revamped. Nowadays significant emphasis is placed on "informed consent" so that participants know that they are involved in a clinical study, what it is hoping to find, and the potential risks. However, these experiments have left a legacy of mistrust within the American black community regarding government health services. For other reasons, this feeling is also prevalent in many parts of Africa, where introducing new vaccines or

EXPORTING DISEASES

The exploitation of poor farmers during the Tuskegee research (see main text) was not an isolated case. In 2005 it was discovered that American scientists had conducted similar experiments in Guatemala between 1946 and 1948. The U.S. and Guatemalan officials organized a study in which soldiers, prisoners, and mental patients were infected with syphilis and other sexually transmitted diseases. Fifteen hundred subjects were involved but no results were ever published. The main person responsible for these experiments was John C. Cutler, who had also participated in the Tuskegee experiment. Cutler died in 2003, and it was only when an archive of his documents was donated to the University of Pittsburgh that the research was uncovered. Medical historian Susan Reverby found reports of the Guatemala experiment while browsing the archive for another study and blew the whistle.

The U.S. government officially apologized to Guatemala in 2010. Also, an international panel of experts was assembled by the United States to investigate current clinical trials being carried out all over the world, ensure that volunteers are treated ethically, and ensure that what happened in Guatemala is never repeated.

treatments for common diseases can be met with frank hostility from the population and even the governments (see box: Exporting Diseases).[16]

LOOKING FOR GUINEA PIGS

Some early biomedical experiments may be conducted on animal species, but eventually any medicine for human use needs to be tested on people before being widely marketed. Currently, clinical trials have to be carefully designed and granted approval by an ethics committee. Initial tests, known as Phase 1, will be carried out with a small group of healthy volunteers to make sure that there are no nasty side effects. Some people become regular participants in these studies (see box: The Birth of the Professional Volunteer).[17]

If no problems come to light, the researchers will be given permission to move on to testing the efficacy of the drug with people that have the relevant illness. This will again involve only a small number of people, perhaps 200 to start with, some of whom may get a mock drug without any of the active ingredient being analyzed, called a **placebo**. The gold standard test is a double-blind study, so called because neither the patients nor the doctors directly involved in running the trial know who is getting the real treatment and who is getting the dummy drug. This process is used to guard against apparent improvement not actually resulting from the medicine

---

THE BIRTH OF THE PROFESSIONAL VOLUNTEER

Until the mid-1970s, "volunteers" for Phase 1 clinical trials were usually prison inmates. Following ethical concerns, rules were changed to open this to anyone and remunerate them for the "inconvenience" of undergoing the experiments (officially, they are paid only for travel expenses and other costs). Depending on the nature of the experiment, volunteers can earn up to $5,000 for only a few weeks' "work." There is actually an anarchist community in Philadelphia that primarily funds itself via this kind of money, thus avoiding in their minds being "exploited" by capitalists.

However, this still raises a problem: a new breed of "professional" guinea pigs has emerged. This is a complex ethical issue: if somebody decides to participate in a trial because they need the money, the whole concept of being a willing volunteer is actually blurred.

but rather from the **placebo effect**, the psychological effects of being involved in an exciting project. The use of placebos actually raises ethical questions itself; patients given the placebo are in effect not being treated with the medicine that may potentially help them, despite the fact that they are suffering with an illness. It is controversial, but because it is important to rule out psychological effects and presently there are no other alternatives, the use of placebos is allowed to continue.

If the new medicine is still looking promising after these initial small trials, it will be tested with a larger group of people, two or three thousand, and possibly compared directly against the best existing treatment for the condition. The study at this stage will deliberately recruit a broader range of people of different ages and racial backgrounds to make sure the drug is still safe for all potential users. Finally, after 10 or more years of research, the new medicine may be licensed for more general prescription by doctors. But often there would be reasons to scratch the whole project, whether it's because the positive effects were not as good as expected or because the side effects were too strong. According to recent studies, each drug that actually reaches the market requires an investment of a billion dollars and more than a million hours of work.

All this costly effort to comply with regulations and follow strict procedures is an indication of how seriously drug development is taken. In fact, a study can be shut down these days simply because the relevant paperwork

---

CONFLICTS OF INTEREST

In the United States, the institutional review boards (IRBs) are the entities that review and ultimately approve all clinical trials with human volunteers. Some IRBs are government-led, but others are private enterprises. This poses an obvious conflict of interest: if a private IRB is too tough, pharmaceutical companies looking for an approval may start taking their business somewhere else. The risk is therefore that these IRBs may just be too "soft" to protect their income.

The government and journalists have come up with a way to avoid this. They periodically submit fake applications to different IRBs for studies that are riddled with ethical problems. If the IRB "fails" to detect them and approves the trial, the government issues a warning. Most private IRBs perform within the expected standards, but some accumulate warnings and continue with their operations.

is not in order. For instance, in February 2011 German anesthesiologist Joachim Boldt was fired when it was discovered that he didn't have the approval of the appropriate board to carry his clinical trials. Around 90 papers were immediately retracted for this reason, and the new treatment protocols he had described in them were suspended pending an investigation on whether the results were correct.[18]

We wish we could say that the problems associated with research misconduct are a thing of the past. Faced with pressures to meet deadlines or captivated by the glory that would accompany being the first to make a landmark discovery, researchers may still be tempted to cut corners. Hopefully, however, tighter regulations and more careful monitoring are bearing fruit. The recent spate of cases may be the result of better policing of research rather than a fall in the integrity of scientists.

IS SCIENTIFIC RESEARCH ETHICAL?

- Several historical scandals show that vulnerable people have been exploited in the name of scientific discovery.
- Scientists may still be tempted to fabricate or falsify data or inappropriately claim a share in work carried out by other people.
- Better monitoring of research and improved training of future scientists should reduce instances of misconduct.
- Accidental mistakes will always be possible, but the majority of research is carried out to the highest possible standards.

Afterword

In this book we have tried to summarize the current state of play in some of the most exciting fields of science and biomedicine. We have allowed ourselves a certain amount of artistic license in the scenarios used to set the ball rolling in each chapter. In the subsequent discussion of the science and ethics, however, we have attempted to be as accurate and balanced as possible—to chart an appropriate path between fact and fiction, and to avoid imposing our point of view.

When the issues we have been discussing are back in the headlines, as they surely will be in the coming months and years, we hope that you will feel better acquainted with the background to the topic and better equipped to discern where the hope has spilled over into hype.

Of course there will be other new developments, ones that we have not discussed specifically in this book, either because there wasn't space or because they will be so fresh that no one has yet seen them coming. If we have done our job properly, then you should have a few more tools to evaluate not only the plausibility of the innovation from a scientific point of view but also the appropriateness for society.

One of the messages we hope has come through is the fact that just because we *can* do something, it doesn't mean that we *should*. The future belongs to all of us, not only to the scientists, and all voices need to be heard. Equally, just because we personally don't like the sound of a new technique doesn't necessarily mean that it will be bad for us. You should now be able

to pause for reflection and consider all the available options before committing to an opinion.

This is only the beginning of a journey into our collective future. We
hope that you have found this first step exciting, but the real fun should
start now. Get ready to walk into the future with your eyes and minds wide
open.

Notes

CHAPTER 1

1. P. Braude, S. Pickering, F. Flinter, and C. M. Ogilvie. "Preimplantation Genetic Diagnosis." *Nature Reviews Genetics* 3 (2002): 941–55.

2. R. Smith. "Baby Born from Embryo Frozen 20 Years Ago." *The Telegraph*, October 10, 2010. Available at http://www.telegraph.co.uk/news/health/news/8053726/Baby-born-from-embryo-frozen-20-years-ago.html.

3. A. Smajdor. "The Ethics of Egg Donation in the Over Fifties." *Menopause International* 14 (2008): 173–77.

4. G. Tremlett and P. Walker. "Spanish Woman Who Gave Birth through IVF at 66 Dies." *The Guardian*, July 15, 2009. Available at http://www.theguardian.com/world/2009/jul/15/spanish-woman-ivf-dies.

5. D.-A. Davis. "The Politics of Reproduction: The Troubling Case of Nadya Suleman and Assisted Reproductive Technology." *Transforming Anthropology* 17 (2009): 105–16.

6. A. Dobuzinskis. "'Octomom' Doctor Loses California Medical License." Reuters, June 1, 2011. Available at http://www.reuters.com/article/2011/06/01/us-octomom-idUSTRE7507TL20110601.

7. S. Katari, N. Turan, M. Bibikova, O. Erinle, R. Chalian, M. Foster, J. P. Gaughan, C. Coutifaris, and C. Sapienza. "DNA Methylation and Gene Expression Differences in Children Conceived In Vitro or In Vivo." *Human Molecular Genetics* 18 (2009): 3769–78.

8. A. Handyside, E. Kontogianni, K. Hardy, and R. Winston. "Pregnancies from Biopsied Human Preimplantation Embryos Sexed by Y-specific DNA Amplification." *Nature* 344 (1990): 768–70.

9. S. Soini. "Preimplantation Genetic Diagnosis (PGD) in Europe: Diversity of Legislation a Challenge to the Community and Its Citizens." *Medicine and Law* 26 (2007): 309–23.

10. T. Severin and E. Kelsey. "Germany Approves Genetic Testing of Human Embryos." Reuters, July 7, 2011. Available at http://in.reuters.com/article/2011 /07/07/us-germany-embryo-vote-idINTRE7664HJ20110707.

11. S.-K. Templeton. "I Will Have Five Babies to Save My Son." *Sunday Times*, September 4, 2011. Available at http://www.thesundaytimes.co.uk/sto/news/uk _news/Society/article768635.ece.

12. J. Harris. *Enhancing Evolution: The Ethical Case for Making Better People.* Princeton, NJ: Princeton University Press, 2007.

13. BBC. "'Designer Baby' Ethics Fear." October 4, 2000. Available at http://news .bbc.co.uk/1/hi/health/955644.stm.

14. BBC. "'Designer Baby' Ban Quashed." April 8, 2003. Available at http://news .bbc.co.uk/1/hi/health/2928655.stm.

15. BBC. "Couple Fight On for Genetically Selected Baby." August 2, 2002. Available at http://news.bbc.co.uk/1/hi/programmes/breakfast/2167608.stm.

16. L. Culley and N. Hudson. "Fertility Tourists or Global Consumers? A Sociological Agenda for Exploring Cross-border Reproductive Travel." *International Journal of Interdisciplinary Social Sciences* 4 (2010): 139–50.

17. BBC. "Brother's Tissue 'Cures' Sick Boy." October 20, 2004. Available at http:// news.bbc.co.uk/1/hi/health/3756556.stm.

18. BBC. "'Designer Baby' Rules Are Relaxed." July 21, 2004. Available at http:// news.bbc.co.uk/1/hi/health/3913053.stm.

19. BBC. "IVF Treatment Fails Baby Choice Couple." March 4, 2001. Available at http://news.bbc.co.uk/1/hi/scotland/1201790.stm.

20. BBC. "Mum Plans to Select Baby's Sex." September 19, 2002. Available at http://news.bbc.co.uk/1/hi/england/2267069.stm.

21. WHO. *Preventing Gender-Based Sex Selection.* Geneva: World Health Organization, 2011. Available at http://apps.who.int/iris/bitstream/10665/44577/1 /9789241501460_eng.pdf

22. M. Hvistendahl. *Unnatural Selection: Choosing Boys over Girls, and the Consequences of a World Full of Men.* New York: PublicAffairs, 2011.

23. S. Camporesi. "Choosing Deafness with Preimplantation Genetic Diagnosis: An Ethical Way to Carry on a Cultural Bloodline?" *Cambridge Quarterly of Healthcare Ethics* 19 (2010): 86–96.

24. S. Baruch, D. Kaufman, and K. L. Hudson. "Genetic Testing of Embryos: Practices and Perspectives of US In Vitro Fertilization Clinics." *Fertility and Sterility* 89 (2008): 1053–58.

25. R. Gray. "Couple Could Win Right to Select Deaf Baby." *The Telegraph*, April 13, 2008. Available at http://www.telegraph.co.uk/news/uknews/1584948/Couples -could-win-right-to-select-deaf-baby.html.

CHAPTER 2

1. J. B. Lee and C. Park. "Verification That Snuppy Is a Clone." *Nature* 440 (2006): E2–E3.

2. R. Spencer. "Pet Dog to Be Cloned for £75,000." *Daily Telegraph*, February 15, 2008. Available at http://www.telegraph.co.uk/news/worldnews/1578796/Pet -dog-to-be-cloned-for-75000.html.

3. T. Shin, D. Kraemer, J. Pryor, L. Liu, J. Rugila, L. Hoew, S. Buck, K. Murphy, L. Lyons, and M. Westhusin. "A Cat Cloned by Nuclear Transplantation." *Nature* 415 (2002): 859.

4. E. Clark. "Online Dog Cloning Auction." *Gizmag*, June 9, 2008. Available at http://www.gizmag.com/best-friends-again-dog-cloning-service/9449.

5. K. H. S. Campbell, J. McWhir, W. A. Ritchie, and I. Wilmut. "Sheep Cloned by Nuclear Transfer from a Cultured Cell Line." *Nature* 380 (1996): 64–66.

6. S. MacVicar. "Doctor Makes Human Cloning Claims." CBS News, April 22, 2009. Available at http://www.cbsnews.com/news/doctor-makes-human-cloning -claims.

7. S. M. M. Damad. "Human Cloning from the Viewpoint of Islamic Fiqh and Ethics." *Asian Bioethics Review* 3 (2011): 342–50.

8. United Nations. "United Nations Declaration on Human Cloning." March 23, 2005. Available at http://www.un.org/en/ga/search/view_doc.asp?symbol=A /RES/59/280.

9. Council of Europe. "Cloning." Available at http://www.coe.int/en/web /bioethics/cloning.

10. Australian Government. Prohibition of Human Cloning Act 2002. Available at https://www.comlaw.gov.au/Details/C2004A01081.

11. M. Hofreiter. "Mammoth Genomics." *Nature* 456 (2008): 330–31; R. E. Green, J. Krause, et al. "A Draft Sequence of the Neandertal Genome." *Science* 328 (2010): 710–22.

12. H. Dixon. "'First Human Head Transplant Now Possible,' Neurosurgeon Claims." *Daily Telegraph*, July 2, 2013. Available at http://www.telegraph.co.uk /news/science/10154240/First-human-head-transplant-now-possible neurosur geon-claims.html.

13. A. D. Goldberg, C. D. Allis, and E. Bernstein. "Epigenetics: A Landscape Takes Shape." *Cell* 128 (2007): 635–38.

14. J. Meikle. "'Twins' Born Five Years Apart." *The Guardian*, January 4, 2012. Available at http://www.theguardian.com/science/2012/jan/04/twins-born-five -years-apart.

15. BioArts International. "BioArts International Ends Cloning Service: Blasts Black-Market Cloners." Press release, September 10, 2009. Available at http://www .bioartsinternational.com/press_release/ba09_10_09.htm.

CHAPTER 3

1. L. Shepherd, R. E. O'Carroll, and E. Ferguson. "An International Comparison of Deceased and Living Organ Donation/Transplant Rates in Opt-In and Opt-Out Systems: A Panel Study." *BMC Medicine* 12 (2014): 131. doi:10.1186/s12916-014-0131-4.

2. Scientific Registry of Transplant Recipients. Table 1.13a: Unadjusted Graft and Patient Survival at 3 Months, 1, 3, 5 & 10Y Survival (%). December 4, 2012. Available at: http://www.srtr.org/annual_Reports/2011/113a_dh.aspx.

3. D. A. Stoffers, T. J. Kieffer, M. A. Hussain, D. J. Drucker, S. Bonner-Weir, J. F. Habener, and J. M. Egan. "Insulinotropic Glucagon-like Peptide 1 Agonists Stimulate Expression of Homeodomain Protein IDX-1 and Increase Islet Size in Mouse Pancreas." *Diabetes* 49 (2000): 741–48.

4. G. R. Martin. "Isolation of a Pluripotent Cell Line from Early Mouse Embryos Cultured in Medium Conditioned by Teratocarcinoma Stem Cells." *Proceedings of the National Academy of Sciences, USA* 78 (1981): 7634–38.

5. J. A. Thomson, J. Itskovitz-Eldor, S. S. Shapiro, M. A. Watnitz, J. J. Swiergiel, V. S. Marshall, and J. M. Jones. "Embryonic Stem Cell Lines Derived from Human Blastocysts." *Science* 282 (1998): 1145–47.

6. M. Wadham. "Court Quashes Stem-Cell Lawsuit." *Nature* 476 (2011): 14–15.

7. D. Dhar and J. H.-e. Ho. "Stem Cell Research Policies around the World." *Yale Journal of Biology and Medicine* 82 (2009): 113–15.

8. A. P. Beltrami, L. Barlucchi, et al. "Adult Cardiac Stem Cells Are Multipotent and Support Myocardial Regeneration." *Cell* 114 (2003): 763–76.

9. J. Kajstura, M. Rota, et al. "Evidence for Human Lung Stem Cells." *New England Journal of Medicine* 364 (2011): 1795–806.

10. A. Giorgetti, N. Montserrat, et al. "Generation of Induced Pluripotent Stem Cells from Human Cord Blood Using OCT4 and SOX2." *Cell Stem Cell* 5 (2009): 353–57.

11. I. Wilmut, N. Beaujean, P. A. de Sousa, A. Dinnyes, T. J. King, L. A. Paterson, D. N. Wells, and L. E. Young. "Somatic Cell Nuclear Transfer." *Nature* 419 (2002): 583–87.

12. K. H. S. Campbell, J. McWhir, W. A. Ritchie, and I. Wilmut. "Sheep Cloned by Nuclear Transfer from a Cultured Cell Line." *Nature* 380 (1996): 64–66.

13. D. Cyranoski. "Woo Suk Hwang Convicted, but Not of Fraud." *Nature* 461 (2009): 1181.

14. K. Takahashi and S. Yamanaka. "Induction of Pluripotent Stem Cells from Mouse Embryonic and Adult Fibroblast Cultures by Defined Factors." *Cell* 126 (2006): 663–76.

15. K. Hayashi, H. Ohta, K. Kurimoto, S. Aramaki, and M. Saitou. "Reconstitution of the Mouse Germ Cell Specification Pathway in Culture by Pluripotent Stem Cells." *Cell* 146 (2011): 519–32.

16. K. J. Brennand, A. Simone, et al. "Modelling Schizophrenia Using Human Induced Pluripotent Stem Cells." *Nature* 473 (2011): 221–25.

17. H.-J. Cho, C.-S. Lee, et al. "Induction of Pluripotent Stem Cells from Adult Somatic Cells by Protein-Based Reprogramming without Genetic Manipulation." *Blood* 116 (2010): 386–95.

18. T. Vierbuchen, A. Ostermeier, Z. P. Pang, Y. Kokubu, T. C. Südhof, and M. Wernig. "Direct Conversion of Fibroblasts to Functional Neurons by Defined Factors." *Nature* 463 (2010): 1035–41.

19. E. Szabo, S. Rampalli, R. M. Risueño, A. Schnerch, R. Mitchell, A. Fiebig-Comyn, M. Levadoux-Martin, and M. Bhatia. "Direct Conversion of Human Fibroblasts to Multilineage Blood Progenitors." *Nature* 468 (2010): 521–26.

20. K. Kim, A. Doi, et al. "Epigenetic Memory in Induced Pluripotent Stem Cells." *Nature* 467 (2010): 285–90.

21. P. Macchiarini, P. Jungebluth, et al. "Clinical Transplantation of a Tissue-Engineered Airway." *Lancet* 372 (2008): 2023–30

22. P. Jungebluth, E. Alici, et al. "Tracheobronchial Transplantation with a Stem-Cell-Seeded Bioartificial Nanocomposite: A Proof-of-Concept Study." *Lancet* 378 (2011): 1997–2004.

23. A. Raya-Rivera, D. R. Esquiliano, J. J. Yoo, E. Lopez-Bayghen, S. Soker, and A. Atala. "Tissue-Engineered Autologous Urethras for Patients Who Need Reconstruction: An Observational Study." *Lancet* 377 (2011): 1175–82.

24. T. Takebe, K. Sekine, et al. "Vascularized and Functional Human Liver from an iPSC-Derived Organ Bud Transplant." *Nature* 499 (2013): 481–84.

25. M. Eiraku, N. Takata, H. Ishibashi, M. Kawada, E. Sakakura, S. Okuda, K. Sekiguchi, T. Adachi, and Y. Sasai. "Self-Organizing Optic-Cup Morphogenesis in Three-Dimensional Culture." *Nature* 472 (2011): 51–56.

26. H. Pilcher. "Green Light for UK Stem-Cell Trial." *Nature News*, January 19, 2009. doi:10.1038/news.2009.41.

27. P. Sims. "Stem Cell Treatment Allows the Blind to See Again." *Daily Mail*, December 23, 2009. Available at: http://www.dailymail.co.uk/news/article-1237721/Revolutionary-surgery-restores-mans-sight-15-years-ammonia-attack.html.

28. D. Cyranoski. "Stem-Cell Therapy Faces More Scrutiny in China." *Nature* 459 (2009): 146–47.

29. T. Ackerman. "FDA Issues Warning to Sugar Land Stem Cell Company." *Houston Chronicle*, October 16, 2012. Available at http://www.chron.com/news/houston-texas/article/FDA-issues-warning-to-local-stem-cell-company-3953961.php/.

30. S. Kleinert and R. Horton. "Retraction—Autologous Myoblasts and Fibroblasts for Treatment of Stress Urinary Incontinence: A Randomised Controlled Trial." *Lancet* 372 (2008): 789–90.

31. S. Watts. "Stem Cell Doctor Robert Trossel Struck Off by GMC." BBC, September 29, 2010. Available at http://www.bbc.co.uk/news/health-11439711.

32. R. Mendick and A. Hall. "Europe's Largest Stem Cell Clinic Shut Down after Death of Baby." *Telegraph*, May 8, 2011. http://www.telegraph.co.uk/news/worldnews/europe/germany/8500233/Europes-largest-stem-cell-clinic-shut-down-after-death-of-baby.html.

CHAPTER 4

1. J. Savulescu, B. Foddy, and M. Clayton. "Why We Should Allow Performance Enhancing Drugs in Sport," *British Journal of Sports Medicine* 38(2004): 666–70.

2. CNN. "Performance Enhancing Drugs in Sports Fast Facts." Last updated December 28, 2015. Available at http://edition.cnn.com/2013/06/06/us/perfor mance-enhancing-drugs-in-sports-fast-facts.

3. M. Slater. "Did This Doctor Give Drugs to Hundreds of Athletes?" BBC, May 1, 2013. Available at http://www.bbc.co.uk/sport/0/cycling/22361185; R. Goldman. "Lance Armstrong Admits to Doping." ABC News, January 17, 2013. Available at http://abcnews.go.com/US/lance-armstrong-confesses-doping/story?id=182 44003.

4. T. D. Noakes. "Tainted Glory—Doping and Athletic Performance." *New England Journal of Medicine* 351 (2004): 847–49.

5. C. Saudan, N. Baume, N. Robinson, L. Avois, P. Mangin, and M. Saugy. "Testosterone and Doping Control." *British Journal of Sports Medicine* 40 (2006): i21–i24.

6. D. G. Jones. "Enhancement: Are Ethicists Excessively Influenced by Baseless Speculations?" *Medical Humanities* 32 (2006): 77–81.

7. M. S. Bahrke and C. E. Yesalis. "Abuse of Anabolic Androgenic Steroids and Related Substances in Sport and Exercise." *Current Opinion in Pharmacology* 4 (2004): 614–20.

8. F. Sjöqvist, M. Garle, and A. Rane. "Use of Doping Agents, Particularly Anabolic Steroids, in Sports and Society." *Lancet* 371 (2008): 1872–82.

9. J. Knight. "Drugs Bust Reveals Athletes' Secret Steroid." *Nature* 425 (2003): 752.

10. G. C. Roberts, Y. Ommundsen, P.-N. Lemyre, and B. W. Miller. "Cheating in Sport." In *Encyclopedia of Applied Psychology*, edited by C. Spielberger, 313–23. Oxford: Elsevier Academic Press, 2004.

11. G. Lippi, M. Franchini, G. L. Salvagno, and G. C. Guidi. "Biochemistry, Physiology and Complications of Blood Doping: Facts and Speculation." *Critical Reviews in Clinical Laboratory Sciences* 43 (2006): 349–91.

12. H. Karp. "Novak Djokovic's Secret: Sitting in a Pressurized Egg." *Wall Street Journal*, August 29, 2011. Available at http://www.wsj.com/articles/SB1000142405 3111904787404576532854267519860.

13. N. Gledhill. "Blood Doping and Related Issues: A Brief Review." *Medicine and Science in Sport and Exercise* 14 (1982): 183–89.

14. L. DeFrancesco. "The Faking of Champions." *Nature Biotechnology* 22 (2004): 1069–71.

15. J. Bonné. "'Go Pills': A War on Drugs?" NBC News, January 9, 2003. Available at http://www.nbcnews.com/id/3071789/ns/us_news-only/t/go-pills-war-drugs /#.Vg-9VY2FMfg; BBC, "'Friendly Fire' Pilots Took 'Go Pills.'" January 15, 2003. Available at http://news.bbc.co.uk/1/hi/world/americas/2657675.stm.

16. V. Cakic. "Smart Drugs for Cognitive Enhancement: Ethical and Pragmatic Considerations in the Era of Cosmetic Neurology." *Journal of Medical Ethics* 35 (2009): 611–15.

17. S. Connor. "Students Could Face Compulsory Drug Tests as Rising Numbers Turn to 'Cognitive Enhancers' to Boost Concentration and Exam Marks." *Independent*, November 7, 2012. Available at http://www.independent.co.uk/news/science/students-could-face-compulsory-drug-tests-as-rising-numbers-turn-to-cognitive-enhancers-to-boost-8289881.html.

18. G. Rousseau. "Giants on Coke." *Nature* 476 (2011): 397.

19. K. R. Urban and W.-J. Gao. "Performance Enhancement at the Cost of Potential Brain Plasticity: Neural Ramifications of Nootropic Drugs in the Healthy Developing Brain." *Frontiers in Systems Neuroscience* 8 (2014): 38. doi:10.3389/fnsys.2014.00038.

20. H. M. E. Azzazy, M. M. H. Mansour, and R. H. Christenson. "Doping in the Recombinant Era: Strategies and Counterstrategies." *Clinical Biochemistry* 38 (2005): 959–65.

21. J. Enriquez and S. Gullans. "Olympics: Genetically Enhanced Olympics Are Coming." *Nature* 487 (2012): 297.

22. J. Kota, C. R. Handy, et al. "Follistatin Gene Delivery Enhances Muscle Growth and Strength in Nonhuman Primates." *Science Translational Medicine* 1 (2009): 6ra15. doi:10.1126/scitranslmed.3000112.

23. L. Battery, A. Solomon, and D. Gould. "Gene Doping: Olympic Genes for Olympic Dreams." *Journal of the Royal Society of Medicine* 104 (2011): 494–500.

24. L. Martinez. "Medal of Honor Awarded to Afghanistan Vet." ABC News, July 12, 2011. Available at http://abcnews.go.com/Politics/medal-honor-awarded-ranger-leroy-petry/story?id=14048891.

25. R. Savill. "Captain Has His Leg Amputated to Stay in Marines." *Telegraph*, May 26, 2004. Available at http://www.telegraph.co.uk/news/uknews/1462841/Captain-has-his-leg-amputated-to-stay-in-Marines.html.

26. BBC. "Amputee Patrick Demonstrates His New Bionic Hand." May 18, 2011. Available at http://www.bbc.co.uk/news/health-13378036.

27. E. Byrne. "Innovation Isn't Safe: The Future According to Kevin Warwick." *Forbes*, September 30, 2013. Available at http://www.forbes.com/sites/netapp/2013/09/30/kevin-warwick-captain-cyborg/.

28. Q. Norton. "A Sixth Sense for a Wired World." *Wired*, July 6, 2006. Available at http://www.wired.com/2006/06/a-sixth-sense-for-a-wired-world.

29. D.-H. Kim, N. Lu, et al. "Epidermal Electronics." *Science* 333 (2011): 838–43.

30. G. Gillett. "Cyborgs and Moral Identity." *Journal of Medical Ethics* 32 (2006): 79–83.

31. N. Bostrom. "In Defence of Posthuman Dignity." *Bioethics* 19 (2005): 202–14.

32. S. Young. *Designer Evolution: A Transhumanist Manifesto*. Amherst, NY: Prometheus Books, 2006.

33. N. Bostrom. "Human Genetic Enhancements: A Transhumanist Perspective." *Journal of Value Inquiry* 37 (2003): 493–506.

34. F. W. Booth, B. S. Tseng, M. Flück, and J. A. Carson. "Molecular and Cellular Adaptation of Muscle in Response to Physical Training." *Acta Physiologica Scandinavia* 162 (1998): 343–50.

35. M. Schuelke, K. R. Wagner, L. E. Stolz, C. Hübner, T. Riebel, W. Kömen, T. Braun, J. F. Tobin, and S.-J. Lee. "Myostatin Mutation Associated with Gross Muscle Hypertrophy in a Child." *New England Journal of Medicine* 350 (2004): 2682–88.

36. C. E. Thomas, A. Ehrhardt, and M. A. Kay. "Progress and Problems with the Use of Viral Vectors for Gene Therapy." *Nature Reviews Genetics* 4 (2003): 346–58.

37. L. Cong, F. A. Ran, D. Cox, S. Lin, R. Barretto, N. Habib, P. D. Hsu, X. Wu, W. Jiang, L. A. Marraffini, and F. Zhang. "Multiplex Genome Engineering Using CRISPR/Cas Systems." *Science* 339 (2013): 819–23.

CHAPTER 5

1. R. E. B. Watson, S. Ogden, L. F. Cotterell, J. J. Bowden, J. Y. Bastrilles, S. P. Long, and C. E. M. Griffiths. "A Cosmetic 'Anti-ageing' Product Improves Photoaged Skin: A Double-Blind, Randomized Controlled Trial." *British Journal of Dermatology* 161 (2009): 419–26.

2. Medicines and Healthcare Products Regulatory Authority. *A Guide to What Is a Medicinal Product*. August 2012. Available at https://www.gov.uk/government /uploads/system/uploads/attachment_data/file/398998/A_guide_to_what_is_a _medicinal_product.pdf.

3. D. Gems and L. Partridge. "Genetics of Longevity in Model Organisms: Debates and Paradigm Shifts." *Annual Reviews of Physiology* 75 (2013): 621–44.

4. M. Leslie. "Are Telomere Tests Ready for Prime Time?" *Science* 332 (2011): 414–15.

5. T. A. Rando. "Stem Cells, Ageing and the Quest for Immortality." *Nature* 441 (2006): 1080–86.

6. K. B. Beckman and B. N. Ames. "The Free Radical Theory of Aging Matures." *Physiological Reviews* 78 (1998): 547–81.

7. P. Kapahi, D. Chen, A. N. Rogers, S. D. Katewa, P. W.-L. Li, E. L. Thomas, and L. Kockel. "With TOR, Less Is More: A Key Role for the Conserved Nutrient-Sensing TOR Pathway in Aging." *Cell Metabolism* 11 (2010): 453–65.

8. K. I. Block, A. C. Koch, M. N. Mead, P. K. Tothy, R. A. Newman, and C. Gyllenhaal. "Impact of Antioxidant Supplementation on Chemotherapeutic Toxicity: A Systematic Review of the Evidence from Randomized Control Trials." *International Journal of Cancer* 123 (2008): 1227–39.

9. D. M. Hirai, S. W. Copp, P. J. Schwagerl, M. D. Haub, D. C. Poole, and T. I. Musch. "Acute Antioxidant Supplementation and Skeletal Muscle Vascular Conductance in Aged Rats: Role of Exercise and Fiber Type." *American Journal of Physiology* 300 (2011): H1536–H1544.

10. A. M. Cuervo. "Autophagy and Aging: Keeping That Old Broom Working." *Trends in Genetics* 24 (2008): 604–12.

11. R. S. Sohal and R. Weindruch. "Oxidative Stress, Caloric Restriction, and Aging." *Science* 273 (1996): 59–63.

12. R. J. Colman, R. M. Anderson, et al. "Caloric Restriction Delays Disease Onset and Mortality in Rhesus Monkeys." *Science* 325 (2009): 201–4.

13. J. A. Mattison, G. S. Roth, et al. "Impact of Caloric Restriction on Health and Survival in Rhesus Monkeys from the NIA Study." *Nature* 489 (2012): 318–22.

14. M. A. Markus and B. J. Morris. "Resveratrol in Prevention and Treatment of Common Clinical Conditions of Aging." *Clinical Interventions in Aging* 3 (2008): 331–39.

15. L. Fontana, L. Partridge, and V. D. Longo. "Extending Healthy Life Span— From Yeast to Humans." *Science* 328 (2010): 321–26.

16. D. E. Harrison, R. Strong, et al. "Rapamycin Fed Late in Life Extends Lifespan in Genetically Heterogeneous Mice." *Nature* 460 (2009): 392–95.

17. S. N. Sehgal. "Sirolimus: Its Discovery, Biological Properties, and Mechanism of Action." *Transplantation Proceedings* 35 (2003): S7–S14.

18. D. Goodkind. "The World Population at 7 Billion." *Random Samplings* [blog], US Census Bureau, October 31, 2011. Available at http://blogs.census.gov/2011 /10/31/the-world-population-at-7-billion.

19. United Nations. "World Population Prospects: 2015 Revision." Available at http://www.un.org/en/development/desa/population/events/other/10/index .shtml.

20. T. Malthus. *An Essay on the Principle of Population.* London: J. Johnson, 1798. Available at http://www.esp.org/books/malthus/population/malthus.pdf.

21. M. Hvistendahl. "Has China Outgrown the One-Child Policy?" *Science* 329 (2010): 1458–61.

22. BBC. "Province Wants Relaxation of China's One-Child Policy." July 11, 2011. Available at http://www.bbc.co.uk/news/world-asia-pacific-14112066.

23. P. McDonald. "Population in the Asian Century." *East Asia Forum*, March 14, 2012. Available at http://www.eastasiaforum.org/2012/03/14/population-in-the -asian-century.

24. United Nations. "The Universal Declaration of Human Rights." 1948. Available at http://www.un.org/en/universal-declaration-human-rights.

25. Organisation for Economic Co-operation and Development. "Statistics on Average Effective Age and Official Age of Retirement in OECD Countries." 2015. Available at http://www.oecd.org/els/emp/ageingandemploymentpolicies-statistic sonaverageeffectiveageofretirement.htm.

CHAPTER 6

1. N. Zagorski. "Profile of Alec J. Jeffreys." *Proceedings of the National Academy of Sciences USA* 103 (2006): 8918–20.

2. E. Pennisi. "ENCODE Project Writes Eulogy for Junk DNA." *Science* 337 (2012): 1159–61.

3. K. Norrgard. "Forensics, DNA Fingerprinting, and CODIS." *Nature Education* 1 (2008): 35.

4. M. A. Jobling and P. Gill. "Encoded Evidence: DNA in Forensic Analysis." *Nature Review Genetics* 5 (2004): 739–51.

5. P. J. Collins, L. K. Hennessy, C. S. Leibelt, R. K. Roby, D. J. Reeder, and P. A. Foxall. "Developmental Validation of a Single-Tube Amplification of the 13 CODIS STR Loci, D2S1338, D19S433, and Amelogenin: The AmpFlSTR Identifiler PCR Amplification Kit." *Journal of Forensic Sciences* 46 (2004): 1265–77.

6. A. A. Noble. "DNA Fingerprinting and Civil Liberties." *Journal of Law, Medicine and Ethics* 34 (2006): 149–52.

7. R. Ford. "How DNA Proved that Mark Dixie Murdered Sally Anne Bowman." London *Times*, May 7, 2009. Available at http://www.thetimes.co.uk/tto/law/article2213878.ece.

8. BBC. "DNA Helps Police 'Solve' 1975 Joan Harrison Murder." BBC News, February 9, 2011. Available at http://www.bbc.co.uk/news/uk-england-lancashire-12396506.

9. P. Cheston and R. Mowling. "Six Years for Lorry Driver's Killing." *Evening Standard*, April 13, 2012. Available at http://www.standard.co.uk/news/six-years-for-lorry-drivers-killing-6939990.html.

10. M. Kayser and P. de Knijff. "Improving Human Forensics through Advances in Genetics, Genomics and Molecular Biology." *Nature Reviews Genetics* 12 (2012): 179–92; E. Murphy. "Legal and Ethical Issues in Forensic DNA Phenotyping." New York University Public Law and Legal Theory Working Papers 415, 2013. Available at http://lsr.nellco.org/nyu_plltwp/415.

11. H. T. Greely, D. P. Riordan, N. A. Garrison, and J. L. Mountain. "Family Ties: The Use of DNA Offender Databases to Catch Offenders' Kin." *Journal of Law, Medicine and Ethics* 34 (2006): 248–62.

12. BBC. "Isle of Wight Rapist Caught by Daughter's DNA." BBC News, March 19, 2010. Available at http://news.bbc.co.uk/1/hi/england/hampshire/8574507.stm.

13. S. Myska. "Twenty Years of DNA Evidence." BBC News, October 9, 2006. Available at http://news.bbc.co.uk/1/hi/uk/6031749.stm.

14. Innocence Project. "The Cases: DNA Exoneree Profiles." Available at http://www.innocenceproject.org/cases-false-imprisonment.

15. T. Paterson. "DNA Blunder Creates Phantom Serial Killer." *The Independent*, March 27, 2009. Available at http://www.independent.co.uk/news/world/europe/dna-blunder-creates-phantom-serial-killer-1655375.html.

16. D. Frumkin, A. Wasserstrom, A. Davidson, and A. Grafit. "Authentication of Forensic DNA Samples." *Forensic Science International: Genetics* 4 (2010): 95–103.

17. M. A. Jobling and P. Gill. "Encoded Evidence: DNA in Forensic Analysis." *Nature Review Genetics* 5 (2004): 739–51.

18. M. Buchanan. "Verdict Raises DNA Evidence Doubt." BBC News, December 20, 2007. Available at http://news.bbc.co.uk/1/hi/northern_ireland/7154189.stm.

19. N. Gilbert. "DNA's Identity Crisis." *Nature* 464 (2010): 347–48.

20. House of Commons, Home Affairs Committee. "The National DNA Database, Eighth Report of Session 2009–10." March 8, 2010. Available at http://www.publications.parliament.uk/pa/cm200910/cmselect/cmhaff/222/222i.pdf.

21. J. Ge, A. Eisenberg, and B. Budowle. "Developing Criteria and Data to Determine Best Options for Expanding the Core CODIS Loci." *Investigative Genetics* 3 (2012): 1.

22. B. Hepple. "Forensic Databases: Implications of the Cases of S and Marper." *Medicine, Science and the Law* 49 (2009): 77–87.

23. R. Williamson and R. Duncan. "DNA Testing for All." *Nature* 418 (2002): 585–86.

CHAPTER 7

1. J. H. Baskin, J. G. Edersheim, and B. H. Price. "Is a Picture Worth a Thousand Words? Neuroimaging in the Courtroom." *American Journal of Law and Medicine* 33 (2007): 239–69.

2. A. Berger. "Magnetic Resonance Imaging." *British Medical Journal* 324 (2002): 35.

3. G. O'Connell, J. De Wilde, J. Haley, K. Shuler, B. Schafer, P. Sandercock, and J. M. Wardlaw. "The Brain, the Science and the Media." *EMBO Reports* 12 (2011): 630–36.

4. A. M. Owen, M. R. Coleman, M. Boly, M. H. Davis, S. Laureys, and J. D. Pickard. "Detecting Awareness in the Vegetative State." *Science* 313 (2006): 1402.

5. B. A. Noah. "Two Conflicts in Context: Lessons from the Schiavo and Bland Cases and the Role of Best Interests Analysis in the United Kingdom." *Hamline Law Review* 36 (2013): 239–64.

6. No Lie MRI. Available at www.noliemri.com.

7. F. Tong and M. S. Pratte. "Decoding Patterns of Human Brain Activity." *Annual Review of Psychology* 63 (2012): 483–509.

8. J. Couzin-Frankel. "Brain Scans Not Acceptable for Detecting Lies, Says Judge." *Science Insider*, June 1, 2010. Available at http://news.sciencemag.org/2010/06/brain-scans-not-acceptable-detecting-lies-says-judge.

9. R. P. Cooper and T. Shallice. "Cognitive Neuroscience: The Troubled Marriage of Cognitive Science and Neuroscience." *Topics in Cognitive Science* 2 (2010): 398–406.

10. S. Batts. "Brain Lesions and Their Implications in Criminal Responsibility." *Behavioral Sciences and the Law* 27 (2009): 261–72.

11. D. Ahuja and B. Singh. "Brain Fingerprinting." *Journal of Engineering and Technology Research* 4 (2012): 98–103.

12. P. S. Applebaum. "Through a Glass Darkly: Functional Neuroimaging Evidence Enters the Courtroom." *Psychiatric Services* 60 (2009): 21–23.

13. O. D. Jones, A. D. Wagner, D. L. Faigman, and M. E. Raichle. "Neuroscientists in Court." *Nature Reviews Neuroscience* 14 (2013): 730–36.

14. G. Gluck. "QEEG Accepted in Death Penalty Trial in *Florida v Nelson*." *Biofeedback* 39 (2011): 74–77.

15. H. G. Brunner, M. Nelen, X. O. Breakefield, H. H. Ropers, and B. A. van Oost. "Abnormal Behaviour Associated with a Point Mutation in the Structural Gene for Monoamine Oxidase A." *Science* 262 (1993): 578–80.

16. A. Caspi, J. McClay, T. E. Moffitt, J. Mill, J. Martin, I. W. Craig, A. Taylor, and R. Poulton. "Role of Genotype in the Cycle of Violence in Maltreated Children." *Science* 297 (2002): 851–54.

17. M. L. Baum. "The Monoamine Oxidase A (MAOA) Genetic Predisposition to Impulsive Violence: Is It Relevant to Criminal Cases?" *Neuroethics* 6 (2013): 287–306.

18. M. Farisco and C. Petrini. "The Impact of Neuroscience and Genetics on the Law: A Recent Italian Case." *Neuroethics* 5 (2012): 317–19.

19. E. Pieri and M. Levitt. "Risky Individuals and the Politics of Genetic Research into Aggressiveness and Violence." *Bioethics* 22 (2008): 509–18.

CHAPTER 8

1. D. G. Gibson, J. I. Glass, et al. "Creation of a Bacterial Cell Controlled by a Chemically Synthesized Genome." *Science* 329 (2010): 52–56.

2. M. Henderson. "A Bug Called Synthia: Biologist Creates New Life." London *Times*, May 21, 2010. Available at http://www.thetimes.co.uk/tto/science/genetics /article2518916.ece.

3. V. Gill. "'Artificial Life' Breakthrough Announced by Scientists." BBC, May 20, 2010. Available at http://www.bbc.co.uk/news/10132762.

4. J. C. Venter. *A Life Decoded: My Genome, My Life*. London: Penguin Paperbacks Science, 2008.

5. N. Wade. "Researchers Say They Created a 'Synthetic Cell.'" *New York Times*, May 20, 2010. Available at http://www.nytimes.com/2010/05/21/science/21cell .html.

6. J. Craig Venter Institute. "First Self-Replicating Synthetic Bacterial Cell." [Press release] May 20, 2010. Available at http://www.jcvi.org/cms/press/press -releases/full-text/article/first-self-replicating-synthetic-bacterial-cell-constructed -by-j-craig-venter-institute-researcher.

7. ETC. "Patenting Pandora's Bug: Goodbye, Dolly. . . Hello, Synthia!" June 7, 2007. Available at http://www.etcgroup.org/content/patenting-pandora%E2%80 %99s-bug-goodbye-dollyhello-synthia.

8. L. P. Wackett. "Engineering Microbes to Produce Biofuels." *Current Opinion in Biotechnology* 22 (2010): 1–6.

9. E. Andrianantoandro, S. Basu, D. K. Karig, and R. Weiss. "Synthetic Biology: New Engineering Rules for an Emerging Discipline." *Molecular Systems Biology* (2006). doi: 10.1038/msb4100073.

10. Biobricks.org.

11. J. Cello, A. V. Paul, and E. Wimmer. "Chemical Synthesis of Poliovirus cDNA: Generation of Infectious Virus in the Absence of Natural Template." *Science* 297 (2002): 1016–18.

12. V. J. J. Martin, D. J. Pitera, S. T. Withers, J. D. Newman, and J. D. Keasling. "Engineering a Mevalonate Pathway in *Escherichia coli* for Production of Terpenoids." *Nature Biotechnology* 21 (2003): 796–802; iGEM. "TNT/RDX Biosensor

and Bioremediator." Available at http://2009.igem.org/Team:Edinburgh/project main/landmines.

13. C. A. Hutchison III, S. N. Peterson, S. R. Gill, R. T. Cline, O. White, C. M. Fraser, H. O. Smith, and J. C. Venter. "Global Transposon Mutagenesis and a Minimal Mycoplasma Genome." *Science* 286 (1999): 2165–69.

14. A. S. Khalil and J. J. Collins. "Synthetic Biology: Applications Come of Age." *Nature Review Genetics* 11 (2010): 367–79.

15. G. Church. "Now Let's Lower Costs." *Nature* 465 (2010): 422.

16. A. E. Goodenough. "Are the Ecological Impacts of Alien Species Misrepresented? A Review of the 'Native Good, Alien Bad' Philosophy." *Community Ecology* 11 (2010): 13–21.

17. Indymedia. "Direct Action: 8 Countries and 8 Days." October 22, 2009. Available at https://www.indymedia.org.uk/en/2009/10/440410.html.

18. T. Douglas and J. Savulescu. "Synthetic Biology and the Ethics of Knowledge." *Journal of Medical Ethics* 36 (2010): 687–93.

19. D. M. Morens, K. Subbarao, and J. K. Taunberger. "Engineering H5N1 Avian Influenza Viruses to Study Human Adaptation." *Nature* 486 (2012): 335–40.

20. J. Randerson. "Revealed: The Lax Laws That Could Allow Assembly of Deadly Virus DNA." *The Guardian*, June 14, 2006. Available at http://www.the guardian.com/world/2006/jun/14/terrorism.topstories3.

21. Presidential Commission for the Study of Bioethical Issues. "New Directions: The Ethics of Synthetic Biology and Emerging Technologies." Available at http://bioethics.gov/synthetic-biology-report.

22. P. Berg. "Asilomar 1975: DNA Modification Secured." *Nature* 455 (2008): 290–91.

23. A. Gutmann. "The Ethics of Synthetic Biology: Guiding Principles for Emerging Technologies." *Hastings Center Report* 41 (2012): 17–22.

24. E. Pennisi. "111 Organizations Call for Synthetic Biology Moratorium." *Science Insider*, March 13, 2012. Available at http://news.sciencemag.org/2012/03/111 -organizations-call-synthetic-biology-moratorium.

25. P. E. M. Purnick and R. Weiss. "The Second Wave of Synthetic Biology: From Modules to Systems." *Nature Reviews Molecular Cell Biology* 10 (2009): 410–22.

26. H. van den Belt. "Playing God in Frankenstein's Footsteps: Synthetic Biology and the Meaning of Life." *Nanoethics* 3 (2009): 257–68.

CHAPTER 9

1. A. E. Shamoo and D. B. Resnik. *Responsible Conduct of Research* (second edition). New York: Oxford University Press, 2009.

2. F. Wells and M. Farthing. *Fraud and Misconduct in Biomedical Research* (fourth edition). London: Royal Society of Medicine Press, 2008.

3. D. Goodstein. *On Fact and Fraud: Cautionary Tales from the Front Lines of Science*. Princeton, NJ: Princeton University Press, 2010.

4. R. Smith. "Research Misconduct: The Poisoning of the Well." *Journal of the Royal Society of Medicine* 99 (2006): 232–37.

5. G. Harris. (2009) "Doctor's Pain Studies Were Fabricated, Hospital Says." *New York Times*, March 10, 2009. Available at http://www.nytimes.com/2009/03/11/health/research/11pain.html.

6. H. C. Sox and D. Rennie. "Research Misconduct, Retraction, and Cleansing of the Medical Literature: Lessons from the Poehlman Case." *Annals of Internal Medicine* 144 (2006): 609–13.

7. BBC. "German Defence Minister Guttenberg Resigns over Thesis." March 1, 2011. Available at http://www.bbc.co.uk/news/world-europe-12608083.

8. BBC. "Media Doctor Admits Plagiarism." June 16, 2008. Available at http://news.bbc.co.uk/1/hi/health/7452877.stm.

9. M. Rossner and K. M. Yamada. "What's in a Picture? The Temptation of Image Manipulation." *Journal of Cell Biology* 166 (2004): 11–15.

10. F. C. Fang, A. Casadevall, and R. P. Morrison. "Retracted Science and the Retraction Index." *Infection and Immunity* 79 (2011): 3855–59.

11. R. G. Steen. "Retractions in the Scientific Literature: Is the Incidence of Research Fraud Increasing?" *Journal of Medical Ethics* 37 (2010): 249–53.

12. A. J. Wakefield, S. H. Murch, et al. "Ileal Lymphoid Nodular Hyperplasia, Non-Specific Colitis, and Pervasive Developmental Disorder in Children." *Lancet* 351 (1998): 637–41 [retracted].

13. B. Deer. "Secrets of the MMR Scare: How the Case against the MMR Vaccine Was Fixed." *British Medical Journal* 342 (2011): c5347.

14. Health Protection Agency. Mumps Notifications (Confirmed Cases), England and Wales. 2012. Available at http://www.hpa.org.uk/web/HPAweb&HPAwebStandard/HPAweb_C/1195733790975.

15. J. H. Jones. *Bad Blood: The Tuskegee Syphilis Experiment* (Revised edition). New York: Free Press, 1993.

16. C. McGreal. "US Says Sorry for 'Outrageous and Abhorrent' Guatemalan Syphilis Tests." *The Guardian*, October 1, 2010. Available at http://www.theguardian.com/world/2010/oct/01/us-apology-guatemala-syphilis-tests.

17. R. Abadie. *The Professional Guinea Pig: Big Pharma and the Risky World of Human Subjects*. Durham, NC: Duke University Press, 2010.

18. J. Wise. "Boldt: The Great Pretender." *British Medical Journal* 346 (2013): f1738.

Index

About the Authors

Chris Willmott, PhD (Biochemistry), MA (Bioethics), is a senior lecturer in the Department of Molecular and Cell Biology at the University of Leicester (UK). His principal interests are the ethics of current innovations in biomedicine and media representations of science. In 2005, Chris was awarded a National Teaching Fellowship. From 2008 to 2015 he served a full term on the education committee for the Nuffield Council on Bioethics. He is the author of *Biological Determinism, Free Will and Moral Responsibility: Insights from Genetics and Neuroscience* (SpringerBriefs Ethics, 2016), an introduction to neurolaw.

Salvador Macip, MD, PhD, is a biomedical researcher and writer. He is a lecturer in the Department of Molecular and Cell Biology at the University of Leicester (UK), where he leads a research group on cancer and aging. He has published 24 books in Spain, some of which have been translated into French and Portuguese, with upcoming translations in Turkish, German, and Italian. His books include fiction, popular science, and books for children and young adults, and he has won several awards. He writes regularly about science for Spanish newspapers.